1,000,000 Books
are available to read at

Forgotten Books

www.ForgottenBooks.com

Read online
Download PDF
Purchase in print

ISBN 978-1-5280-2557-7
PIBN 10898735

This book is a reproduction of an important historical work. Forgotten Books uses state-of-the-art technology to digitally reconstruct the work, preserving the original format whilst repairing imperfections present in the aged copy. In rare cases, an imperfection in the original, such as a blemish or missing page, may be replicated in our edition. We do, however, repair the vast majority of imperfections successfully; any imperfections that remain are intentionally left to preserve the state of such historical works.

Forgotten Books is a registered trademark of FB &c Ltd.
Copyright © 2018 FB &c Ltd.
FB &c Ltd, Dalton House, 60 Windsor Avenue, London, SW19 2RR.
Company number 08720141. Registered in England and Wales.

For support please visit www.forgottenbooks.com

1 MONTH OF FREE READING

at

www.ForgottenBooks.com

By purchasing this book you are eligible for one month membership to ForgottenBooks.com, giving you unlimited access to our entire collection of over 1,000,000 titles via our web site and mobile apps.

To claim your free month visit:
www.forgottenbooks.com/free898735

* Offer is valid for 45 days from date of purchase. Terms and conditions apply.

English
Français
Deutsche
Italiano
Español
Português

www.forgottenbooks.com

Mythology Photography **Fiction** Fishing Christianity **Art** Cooking Essays Buddhism Freemasonry Medicine **Biology** Music **Ancient Egypt** Evolution Carpentry Physics Dance Geology **Mathematics** Fitness Shakespeare **Folklore** Yoga Marketing **Confidence** Immortality Biographies Poetry **Psychology** Witchcraft Electronics Chemistry History **Law** Accounting **Philosophy** Anthropology Alchemy Drama Quantum Mechanics Atheism Sexual Health **Ancient History Entrepreneurship** Languages Sport Paleontology Needlework Islam **Metaphysics** Investment Archaeology Parenting Statistics Criminology **Motivational**

PREFACE

'estock auction market in the Southeast is an important institution. During the
les there has been a rapid increase in the number of these markets. It was almost
.hat many of the facilities would be improperly designed and arranged for most
ıerations. Today, at many auctions, labor costs are excessive.

;earch which is the basis for this report was undertaken to provide data and
/hich can be used in establishing and operating auction market facilities more
The study was conducted in cooperation with the Departments of Agricultural
ıd Sociology of the Agricultural Experiment Stations of Florida, Georgia,
ınd Mississippi.

is due W. K. McPherson of the Florida Experiment Station; J. T. Harris, formerly
ırgia Experiment Station; M. D. Woodin and Henry J. Casso of the Louisiana
itation; and W. E. Christian, Jr., and L. Dow Welch of the Mississippi Experiment
their assistance in the study.

hors wish to thank the market operators who made their facilities available for
.de suggestions for increasing the efficiency of market operations and facilities.

acknowledgment is made to A. F. Schramm, industrial engineer, Packers and
ranch, Livestock Division, Agricultural Marketing Service, for his assistance on
oblems of facility design, market layouts, ınd construction costs.

hors wish to acknowledge use of material on facility designs contained in
Handbook No. 36, *Suggestions for Improving Services and Facilities at Public
·ckyards*.

· Superintendent of Documents, U. S. Government Printing Office, Washington 25, D. C. · Price 50 cents

CONTENTS

	Page
Summary	iii
Introduction	1
Types of livestock auction markets in the Southeast	2
Type 1. Markets that weigh livestock following their sale	3
Type 2. Markets that weigh and sort hogs on arrival and weigh cattle before their sale	10
Type 3. Markets that weigh livestock before their sale	16
Description and defects of market facilities in the Southeast	19
Yard facilities	19
Sales barns	33
Arrangement of facilities	35
Summary of defects	40
Suggested designs for livestock market facilities in the Southeast	42
Truck docks	42
Cattle tagging chute	47
Cattle fences and gates	48
Hog fences and gates	49
Scale platform and rack	50
Feeder-chute facility	52
Cattle trough	53
Hog trough	53
Hayrack and grain trough	53
Catwalks	55
Auction barn	55
Suggested layout for a market that weighs livestock after their sale--type 1	59
Facilities needed	59
Arrangement of facilities	66
Estimated cost of construction	69
Amount of land needed for market site	69
How suggested facility should operate	69
Suggested layout for a market that sorts and weighs hogs on arrival and weighs cattle before their sale--type 2	74
Facilities needed	74
Arrangement of facilities	80
Amount of land needed for market site	80
Estimated cost of construction	82
How suggested facility should operate	83
Suggested layouts for markets that weigh livestock before their sale--type 3	87
Layout for market that tags but does not dip cattle	87
Layout for market that dips but does not tag cattle	87
Potential benefits from improved livestock auction market facilities	99

SUMMARY

Based on key practices and the kinds of livestock handled, auction markets in the Southeast are of 3 general types: (1) Markets that weigh livestock after their sale, (2) markets that sort and weigh hogs on arrival and weigh cattle before their sale, and (3) markets that weigh livestock before their sale. The last type is further divided into markets that: (a) Tag cattle but do not dip them, and (b) dip cattle but do not tag them. Each of the 3 types of markets vary in the kinds, amount, and arrangement of facilities. This study was designed to develop an efficient layout for each one of the three.

The principal defects of many livestock auction market facilities, regardless of the type, are: (1) Improperly designed truck docks, (2) inadequate and improperly arranged catch pens, (3) lack of properly designed tagging chutes, (4) holding pens too large, (5) improper type floors for holding pens and alleys, (6) excessive fence and gate construction, (7) improperly located scales, (8) lack of properly designed facilities for driving livestock into the sales ring, (9) improperly arranged yard facilities for a direct flow of livestock through market operations, (10) improperly designed sales rings, (11) poorly designed sales barns, and (12) small market sites.

Livestock are moved through all 3 types of markets in 3 cycles or groups of operations: (1) Receiving, (2) selling, and (3) loading out. The operations in each cycle are usually performed in regular sequence--animals are driven directly from one operation to another--but there are variations in these sequences. Most auctions of all types have work crews large enough to take care of a volume of business near the upper limits handled on peak sales. Therefore, on some days there may be little relationship between the workload and the crew size. Of course, the kind of facilities used and their arrangement affect the number of workers required.

The number of workers, other than office personnel, usually employed by typical markets on sales days are: (1) Type 1 markets that handle about 600 cattle and 150 hogs per sale, 22 workers; (2) type 2 markets that handle about 1,000 hogs and 200 cattle per sale, 18 workers; and (3) type 3 markets that handle about 600 cattle and 50 hogs per sale, 23 workers.

Suggested layouts for each of the 3 types of markets include improved designs and arrangements of yards, auction barn, market driveway, and parking areas for motor vehicles. Yard facilities include truck docks, tagging chute, sorting pen, catch pens, holding pens for cattle and hogs, alleys, scales, scale pockets, feeder chute, loading pens, catwalks, and dipping vat. The kinds and amounts of facilities needed and their arrangement vary considerably for each of the 3 types of markets. The same auction barn design is suggested, with slight modifications, for all 3 types of markets. It is 66 feet wide and 45 feet deep plus an annex for the auctioneer's box. The barn proper contains space for the auctioneer's box, sales ring, seating area, market offices, restaurant, lobby, and toilet facilities. A 40-foot main driveway connecting with 100-foot aprons or service driveways and parking space for about 210 motor vehicles are suggested.

The estimated costs for constructing each of the 3 types of market facilities in 1954, in line with the suggested layouts are, roughly: Type 1, $69,500; type 2, $70,500; type 3, $73,800. To the cost of construction should be added the cost of land placed in condition to build. A minimum of 7.5 acres of land is needed for any of the 3 types of markets.

It is estimated that the suggested markets could be operated on sales days with 3 fewer workers than are now required on typical markets of comparable size or handling a comparable volume of livestock. In addition, some reduction in labor required for checking animals in buyer pens after the sale should be possible. Other benefits possible from properly designed and arranged facilities are a reduction in losses from shrinkage, bruises and injuries, and deaths. The benefits from these items probably would be as great as, if not greater than, the benefits derived from operating with smaller crews. Improved facilities also should (1) provide sanitary conditions which would assist in preventing the spread of livestock diseases, (2) reduce the amount of time sellers' trucks spend waiting to unload, (3) increase the rate at which buyer trucks can be loaded, and (4) assure an uninterrupted flow of livestock into the sales ring.

LIVESTOCK AUCTION MARKETS IN THE SOUTHEAST -
METHODS AND FACILITIES

By George E. Turner, agricultural economist, and
Clayton F. Brasington, industrial engineer,
Transportation and Facilities Branch,
Marketing Research Division
Agricultural Marketing Service

INTRODUCTION

auction markets are comparatively new in the marketing of livestock in the
rn States. Although such markets appeared in Georgia as early as about 1925, there
few in the Southeast before 1930. The first Mississippi auction for which records
ble began operating in 1932. The first auction in Louisiana for selling cattle,
d hogs was organized in 1934. By 1937 only 83 auctions were reported to be operating
ippi, Georgia, Florida, Louisiana, Alabama, North Carolina, and South Carolina. In
umber of auctions in these States had increased to 392.

auction markets are the most important agency in the marketing of livestock in the
ith minor exceptions, more than half of every class of livestock marketed is sold at
market. These markets serve the largest number of producers and also handle the
lume of livestock.

few exceptions, auction markets in the Southeast sell all species of livestock.
he number of horses, mules, goats, and sheep handled is comparatively small. Cattle
re handled by at least 95 percent of the markets. Only a few auctions sell cattle
gs only.

but few exceptions, auction markets in the Southeast conduct only one sale a week.
quence, market facilities are not fully utilized, which is in contrast with most
nesses that operate 5 or 6 days each week. In some instances auctions accept live-
igned for sale a day or so prior to the sale, or hold buyers' livestock for short
ter the sale. In other cases auctions act as buying stations for certain species of
throughout the week. However, the portion of the facilities used for these activities
tively small and auctions employing these practices are the exception rather than

olume of business done by livestock auction markets in the Southeast varies widely
to sale. Part of this variation is the result of seasonal fluctuations. However,
onditions and community events are important factors in the variation of the volume

of business. Variations of as much as eight times the low volume are common for cattle and even larger variations are common for hogs. Auction operators cannot always predict accurately the volume that may be offered for a particular sale.

Most markets tend to build their facilities of a size adequate to accommodate the upper limits of their sales volume. As a result, for a number of sales during the year the pen space utilization is low. However, during the few peak sales the pens are crowded with livestock.

As should be expected from a business operating only 1 day a week and with an unpredictable volume of business, the maintenance of a trained labor force is a major problem. In addition to yard workers, this problem also involves office workers. Many markets employ farmers or schoolboys part time as yardmen. Most markets have a high labor turnover, particularly of yardmen, and new workers have to be trained to perform efficiently. Some markets guarantee their workers a full day's wage regardless of the length of the sale. Some markets either pay their workers by the hour or pay a guaranteed minimum with additional wages for each hour of work above the prescribed minimum.

Physical operations on sales days are divided into 3 major cycles: (1) Receiving livestock, (2) selling livestock, and (3) loading livestock. Although each of these cycles usually represents a separate time period during the day, nearly all markets have short overlapping cycles.

Because of their rapid growth, a relatively large number of Southeastern auction markets do not have facilities and equipment for the most efficient handling and selling of livestock. Many of the facilities are improperly designed and arranged and, as a result, require excessive amounts of labor. Moreover, the rough handling of livestock--due in part to the design of the facilities--results in an excessive amount of bruising, injuries, and shrinkage.

The purpose of the research reported here was: (1) To design improved livestock auction market facilities, (2) to determine the extent to which established market practices affect facility layouts and designs, and (3) to develop more efficient methods for receiving, selling and loading livestock at auction markets.

In this research 20 markets were selected for case study in different geographical areas. Five auctions were selected in each of the following States: Mississippi, Florida, Georgia, and Louisiana. Data were obtained through time studies and analyses of methods at each market. In addition, a layout and flow diagram of several of the facilities was prepared. Selected operations and the facilities at a number of other markets also were studied.

TYPES OF LIVESTOCK AUCTION MARKETS IN THE SOUTHEAST

Weighing and sorting different species of livestock are considered to be key practices as they influence more than others the layout, designs, and operations of auctions. Based on these key practices the markets in the Southeast are divided into 3 general types: (1) Markets that weigh livestock immediately after they are sold, (2) markets that weigh and sort hogs into market grades on arrival and weigh cattle immediately before they are sold, and (3) markets that weigh livestock immediately before they are sold. These 3 types of markets do not include all variations and combinations of market practices in the region. However, these key practices are ones which must be considered in designing improved facilities.

There is a high degree of similarity in the operations of all 3 types of markets. Placing tags on cattle in markets that weigh livestock after their sale may be performed in the same manner on markets that weigh livestock before their sale. The weighing operation is similar on all 3 types of markets, although the sequence in which the operation is performed differs between 2 types of markets. Again certain operations are performed in only one type of market. Hogs are sorted into *grades* and weight classes only at markets that weigh and sort hogs on arrival and weigh cattle before their sale. *Grades* are usually groupings determined by sex and condition in addition to weight. Most markets selling relatively large numbers of hogs sort hogs into about 10 *grades*.

All States have statutes designed to prevent the spread of livestock diseases. Some of these statutes affect the layout of auction facilities. Since statutes are changed from time to time, anyone intending to build new facilities or remodel old ones should check the requirements of his own State before developing plans.

Type 1. Markets That Weigh Livestock Following Their Sale

In addition to weighing livestock as they leave the sales ring, many markets tag and mix pen animals, provide both buyer and seller with holding pens, start the bidding, and sell most animals singly. Most of these markets receive and ship nearly all livestock by motortrucks. This type of market is the most prevalent kind in Louisiana and Mississippi. Figure 1 shows the layout of an auction market that practices after-sale weighing of livestock. A market of this type handles about 600 cattle and 150 hogs per sale.

Receiving Livestock

Cattle are unloaded at the truck dock, if necessary separated by ownership in the chute pen, tagged, the receiving ticket prepared, assigned to sellers' holding pens, and yarded in the sellers' holding pens. With the exception of tagging, the same operations are performed in receiving hogs. Many markets with this volume of business receive cattle and hogs at the same truck docks. One worker can usually perform all the receiving operations for hogs, and frequently in conjunction with receiving cattle. Crew sizes for receiving ranged in size from 8 to 14 workers. The workers preparing the receiving ticket and those tagging cattle usually have no other duties. Other members of the crew usually perform various receiving operations as required.

Two major methods are used in receiving cattle. By one method all cattle arriving at the market are driven into the tagging chute for tagging. By the other method cattle arriving in small trucks are tagged before they are unloaded and cattle arriving on large trucks are unloaded and driven into the tagging chute for tagging. Table 1 shows the hourly rate at which cattle and hogs were received by each of the 2 methods. The time required to receive a truckload of hogs is about the same as that required for cattle. The volume of receipts was large enough to provide the crews a full and steady workload.

When cattle were tagged both in the tagging chute and at the truck docks, 51 truckloads were received per hour by a 9-man crew, or an average of 5.7 truckloads per man-hour. When all cattle were tagged in the tagging chute, 29 truckloads were received per hour by an 8-man crew, an average of 3.6 truckloads per man-hour. The number of large truckloads of cattle received per hour is about the same for both methods. However, the number of small truckloads

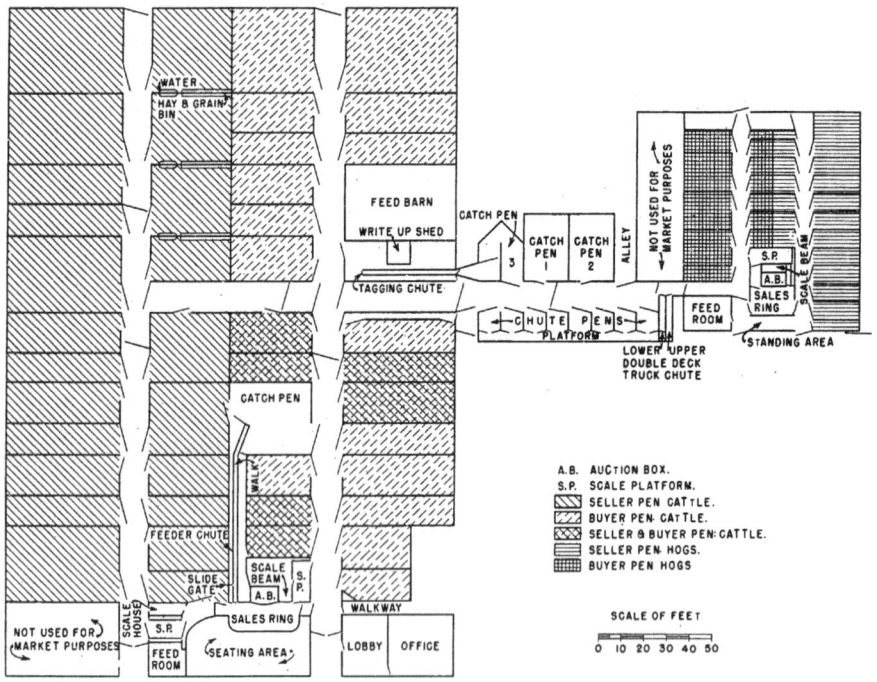

Figure 1.--The layout of an auction that practices after-sale weighing of livestock.

of cattle received per hour when cattle are tagged on the trucks is almost twice as great as when the cattle are tagged in the tagging chute. The difference stems from the fact that on small trucks, the tagging and writeup operations are performed while the tailgate of the truck is being detached, permitting 3 operations to be carried on simultaneously--unloading, tagging, and preparing the ticket. Although an additional worker is employed when cattle are tagged both at the truck dock and in the tagging chute, the number of truckloads received per manhour is considerably higher than when all cattle are driven to the tagging chute.

These data indicate that auctions receiving a large part of their cattle on pickup trucks could expedite receiving operations by tagging cattle on these trucks and driving cattle arriving on large trucks into the tagging chute. This method should enable auctions to receive cattle as they arrive and help prevent trucks from lining up outside the market.

received per hour when all cattle are tagged in the tagging
le are tagged in the tagging chute and at the truck dock

Animals received				Total		Men in:	Truck-
small	: On large	:	:	:	:receiv-:	loads
rucks	: trucks	:Trucks	:Cattle:	Hogs	: ing	:per man-
le : Hogs	:Cattle: Hogs	:	:	:	: crew	: hour
ber Number	Number Number	:Number	Number	Number	:Number :	Number
7 22	102 13	: 29	149	35	: 8 :	3.6
9 17	101 3	: 51	200	20	: 9 :	5.7

ily pickup trucks.
type body farm trucks or tractor-trailer trucks.
re received at the same truck docks. The time required for
s about the same as that required for cattle.

Selling Livestock

nilar for both cattle and hogs. Animals are brought up from
sales ring, driven singly or in small groups into the sales
ing and sold, driven from the ring onto the scale platform and
ed in buyer pens. All operations in the selling cycle require a
er, the number of workers assigned to specific operations varies
e size of the crews used in selling cattle ranged from 16 to
elling hogs ranged from 13 to 20 workers. The number of workers
ly fewer because the area of the hog division is considerably
re shorter.

rings up cattle for sale. One worker performs the same operation
s use two workers because the layout necessitates long out-of-
from crossflows.

icilities used for driving cattle into the sales ring are the
long, narrow chute. Four or 5 workers are required to drive
1 these facilities. Cattle may be driven into the sales ring
and feeder chute facility with 3 workers, even in markets
ely fast rate. The number of workers used in driving hogs into
ies of small catch pens ranged from 2 to 6 workers. The multiple
on auctions of this type in Texas, is not being used in the

Sales ring activities require 2 or 3 ringmen, a starter, an auctioneer, and a recorder--
a total of 5 or 6 workers. Markets having relatively large sales rings and selling at a fast
rate usually employ 3 ringmen. Markets that sell at a comparatively slow rate often perform
the selling with only 2 ringmen, regardless of the size of the ring (fig. 2). However, some
markets that have comparatively small rings sell at a very fast rate with only 2 ringmen.

The number of workers employed in weighing livestock depends on the location of the scale
and the method of driving animals off the scale platform. Where the scale platform is adjacent
to the sales ring and the exit gate of the sales ring is the entrance gate of the platform, a
ringman drives the animals onto the scale platform. With this arrangement only 2 workers are
needed--the weighmaster and the scaleman who drives the animal from the scale platform. However, some markets have difficulty in driving animals from the scale platform fast enough to
keep the weighing operation abreast of the selling. To alleviate this condition, a number of
markets have a side gate cut into the scale platform rack. A worker stationed at this gate

NEG. BN-3146

*Figure 2.--Selling cattle on a livestock auction market in the Southeast. Two ringmen perform
the necessary work.*

enters the scale platform after weighing is completed and drives the animal out (fig. 3). This method minimized delays in the scale, but required 3 workers--2 scalemen and a weighmaster.

NEG. BN-3147

Figure 3.--Driving cattle from the scale platform after weighing.

Where the scale platform is separated from the sales ring, one worker must open the *in* gate of the platform and another worker, the *out* gate. Therefore, under this arrangement, 3 workers are required regardless of speed of the sale. If markets having this arrangement desire to sell at a comparatively fast rate a fourth worker is stationed at a side gate to assist in driving the animals from the platform.

Assigning cattle to buyer pens after weighing usually is performed by the weighmaster. The weighmaster calls over a speaker system to the workers yarding the animals the number of the pen to which the animal is assigned. Although this method is fairly effective, errors frequently are made which necessitate locating mispenned animals. By another method the weighmaster calls both the number of the pen and the number of the tag on the animal over the loudspeaker. Workers then yard the animal in the buyer pen, and write the tag number on a card attached to the pen. This method permits mispenned animals to be located quickly. Estimated labor requirements for locating animals by this method are about half that required by the first method.

As hogs are not tagged, they are yarded in buyer pens by the first method. So few hogs are handled by markets of this type that problems in locating mispenned animals were minor.

Yarding crews range in size from 3 to 6 workers. The size of crews is determined primarily by the arrangement of yard facilities and the layout of the market, rather than by the speed of the sale. However, markets selling animals at a comparatively slow rate tend to use smaller crews than markets selling animals at a fast rate. Where the pen-back alley is at a right angle to the *off* end of the scale or the arrangement of the yards is such that 2 or 3 alleys are used for driving animals to buyer pens, the number of workers employed is larger than where only one drive alley leading straight away from the scale platform is used. Better organized markets usually employ a crew of 4 workers. Typically, the worker adjacent to the *off* end of the scale platform yards cattle in the first 4 pens, the second worker yards in the next 6 pens, the third worker yards in the next 8 pens, and the fourth worker yards in the last 12 pens, for a total of 30 pens. Cattle of the buyers of larger lots of cattle are yarded in pens nearest to the sales ring to equalize the workload of crew members. (Fig. 4.)

Hogs are yarded in buyer pens by most markets by the same method used for cattle. Crews ranged in size from 2 to 5 workers. The smaller crews are used when the hog division is near the sales barn.

Loading-out Livestock

The cycle of loading-out operations involves (1) checking the tag number of the animals against the buyer's sheet, (2) driving the livestock from buyers' pens to the loading dock, and (3) loading onto trucks.

Loading-out usually begins shortly after the start of the sale. While the sale is underway 2 workers usually perform all load-out operations. These 2 workers also receive livestock after the sale begins. After the sale about 10 additional workers usually are assigned to loading-out operations. There usually is not clear division of responsibility for the 3 load-out operations.

Figure 4.--Yarding cattle in a buyer pen.

Summary of Labor Requirements

A typical market that practices after-sale weighing of livestock (type 1) begins receiving livestock about 7 a.m. and continues until just before the sale is over. The sale begins at 12 noon and is completed about 5 p.m. Checking animals for loading out is completed about 7 p.m. Livestock are driven from buyer pens to the loading-out docks throughout the night as buyer trucks arrive. The total number of workers, other than office workers, employed by the typical market is 22. This market has a yard arrangement about average and sells animals at a comparatively fast rate. For the markets studied in this group, the crew number varied from 19 to 25.

The crew of the typical market consists of the auctioneer, starter, weighmaster, recorder, 2 ringmen, the yard foreman, and 15 yardmen. The auctioneer, starter, weighmaster, and recorder usually perform only one operation each during the course of the sale. The yardmen and ringmen perform as many as 2 or 3 different operations. The yard foreman directs the activities of the yard workers and frequently collaborates with them in performing specific jobs.

As receiving begins before the sale and is continued during the sale as additional receipts arrive, receiving and selling operations are conducted concurrently. Loading out may begin shortly after the sale starts, in which case all 3 cycles of operation are performed concurrently. However, receiving and loading-out activities during the sale usually are relatively light and workers are shifted from one job to another.

Ten workers directed by the yard foreman receive livestock before the sale starts. Only the workers preparing the receiving ticket and tagging are *fixed* to their jobs. Other workers in the crew exchange jobs and collaborate with one another. When the hog sale starts, 6 workers are transferred to selling operations. Four workers remain in the receiving crew while hogs are being sold. When the hog sale is over 2 additional workers from the receiving crew are transferred to cattle-selling operations. The 2 remaining workers perform all receiving and loading-out operations during the sale.

The height of activity occurs during the sale. The 17 workers used in selling hogs are obtained by having 6 workers shift from receiving and by having 11 additional workers report for duty. For the cattle sale 19 workers are used--they are the 11 workers which reported at the start of the hog sale and 8 workers from the receiving crew. Most workers are assigned specific jobs which they perform continually throughout the sale. The sale of hogs rarely lasts more than three-quarters of an hour. The number of workers needed for selling cattle influences to some extent the number of workers used for selling hogs. The hog sale usually is of such short duration that some yardmen move into position whether they are needed or not, so that they will be ready for the sale of cattle. The 19 workers selling cattle include 1 worker bringing up livestock for sale, 4 workers driving livestock into the sales ring, 2 ringmen, the auctioneer, recorder, starter, weighmaster, 3 scalemen, and 5 workers yarding livestock in buyer pens. The yard foreman directs the activities of the yardmen performing jobs in the selling operations and also the workers receiving and loading-out livestock during the sale.

At the end of the sale 10 of the yardmen are shifted to loading-out operations. Thus the total crew for this work for a comparatively short period is 12 workers. The principal duties of these workers are to check animals in buyers' pens against the buyers' invoices and to drive the animals from the pen to the loading-out docks. One or two workers remain on duty throughout the night to drive animals from the pens to the loading docks as buyers' trucks arrive on the market.

Type 2. Markets that Weigh and Sort Hogs on Arrival and Weigh Cattle Before their Sale

In addition to sorting hogs by market grades and presale weighing of cattle, many markets tag and mix pen cattle receipts, provide both buyer and seller holding pens, sell a high proportion of cattle singly, and sell hogs in both small and large size lots. These markets receive and ship nearly all livestock by trucks. This type of market is most common in south

- 11 -

and north Florida. A typical market of this type for which the methods of moving
:k are described handles about 1,000 hogs and 200 cattle per sale. Figure 5 shows the
of this type of auction.

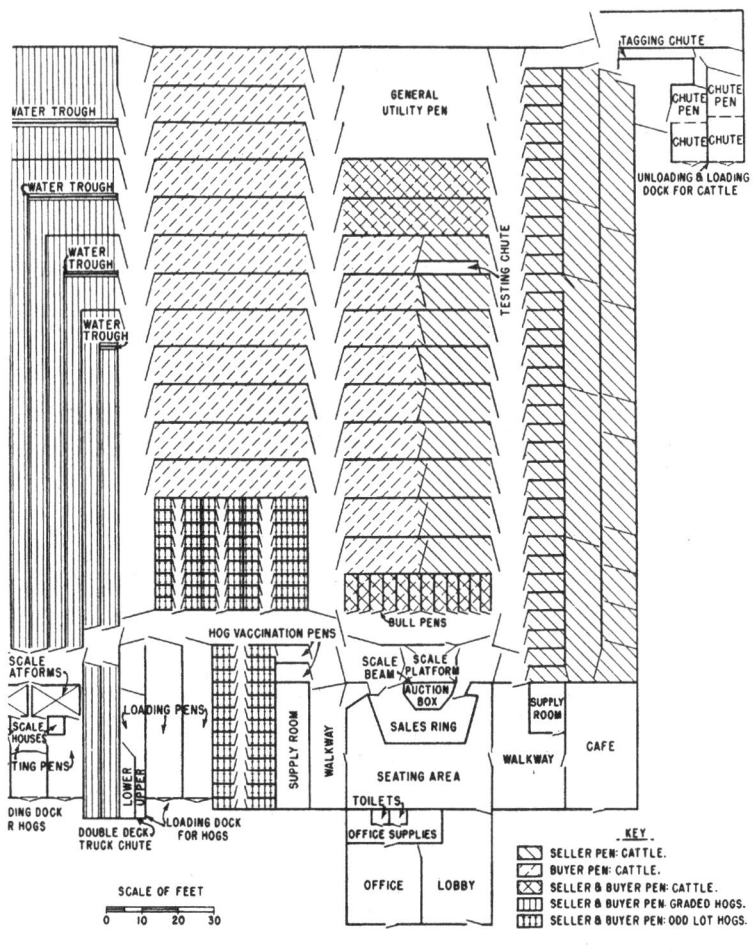

Figure 5.--The layout of an auction that practices sorting hogs and presale weighing of cattle.

Receiving Livestock

Hogs are unloaded from motortruck and, if necessary, separated by ownership in the chute pen. From the chute pen, hogs are driven to the sorting pens where they are sorted by *grades*, weighed, the receiving ticket prepared, and assignments to sellers' holding pens made. Hogs are then driven to sellers' holding pens in the yards. Workers who sort hogs usually drive them onto the scale platform. Preparing the receiving ticket and assigning hogs to seller holding pens usually is done by the weighmaster. Other workers in the receiving crew drive hogs from the scale platform to sellers' pens.

Some markets that sort hogs use 1 scale and some use 2 scales in receiving hogs. Table 2 shows comparative receiving rates with the 2 methods.

Table 2.--Comparative rates of receiving hogs at selected markets that use 1 scale and markets that use 2 scales

Method	Market 1				Market 2				Market 3			
	Trucks 1/	Hogs	Workers in receiving crew	Truck-loads per man-hour	Trucks 1/	Hogs	Workers in receiving crew	Truck-loads per man-hour	Trucks 1/	Hogs	Workers in receiving crew	Truck-loads per man-hour
	Number	Number	Number	Number	Number	Number	Number	Number	Number	Number	Number	Number
Receiving hogs, using 1 scale......	26	224	7	3.7	26	109	7	3.7	24	180	7	3.5
Receiving hogs, using 2 scales.....	37	252	10	3.7	0	0	0	0	0	0	0	0

1/ Trucks include pickup trucks, stake-type body farm trucks, and tractor-trailer trucks.

Markets receiving by the 1-scale method employ 7 workers in the crew. The results of these observations indicate that although the number of truckloads of hogs received by each market per hour is about the same, the number of hogs received per hour varied considerably, ranging from 109 to 224. Thus, it appears that the size of the load has but little effect on the time required to receive a truckload.

When 2 scales are used the number of trucks received per hour with a crew of 10 workers was 37 and the number of hogs received per hour was 252. The results of this observation indicate that the receiving rate would not be twice as great as that when 1 scale is used. The second scale usually is used for only a few days each year and for only short periods on those days.

Cattle receiving operations at markets of this type are the same as those described for type 1 markets. However, as comparatively few cattle are handled, 2 to 5 workers can receive them. A considerable part of the time the crew is idle, regardless of its size. As these

- 13 -

:ive hogs at one set of docks and cattle at another set, 2 receiving crews
number of workers in the combined hog and cattle receiving crews varies
markets, ranging from 9 to 14 workers.

Selling Livestock

for markets of this type range in size from 10 to 22 workers for cattle and
these markets sell hogs in 2 ways--sorted into *grades* and in odd-lots.
l in the sales barn, and are not driven into the sales ring. Only 2 workers
: graded hogs--the auctioneer and the recorder.

ire brought up from the seller's pen to a catch pen near the sales ring,
orkers. Then they are driven through a series of small catch pens into the
it may be sorted into salable lots and each salable lot driven into the
mally 2 or 3 workers drive the hogs into the sales ring.

e ring is usually conducted by the auctioneer, the recorder, and 2 or

ided hogs are sold they are assigned to a buyer pen by a special worker.
·s used for yarding hogs varies from 2 to 5.

i of this type a yardman will enter the pen, drive the hogs out and through
iles ring. He waits for them to be sold, then drives them back to the same
.his method usually have 7 or 8 yardmen to maintain a continuous sale. An
s driven into the sales ring at one time without regard to weight classes
: auctioneer designates the salable hogs of the lot and solicits bids on
ices 3 or 4 sales transactions may be conducted for the lot. Buyers may be-
it be certain which animals they are bidding on. Sometimes transactions are
igs must be resold. Sales are more orderly when all the animals in the sales
:e.

iens are used for both sellers' and buyers' hogs, buyers' hogs usually must
from several pens. Therefore, the use of separate seller and buyer pens
fuired to load-out hogs.

:tle for sale usually is done by one worker. The cattle pens usually are
:e from the sales ring. Driving cattle into the sales ring includes weighing
m serves as a pen through which cattle are driven into the sales ring. A
:rs usually is required to drive cattle through these facilities and into
.ngman opens the *out* gate of the scale platform and drives the cattle into
fuently he is assisted by one of the crew at the rear of the scale platform.

iys occur during sales because animals are not driven from the scale
as they are needed in the ring. Some markets station a worker on the plat-
; into the sales ring more quickly. However, he is more subject to injury
i than a man located outside. Furthermore, weighing livestock with a man on
the possibilities of weighing errors. To move animals more quickly a gate
i figure 6 could be cut into the side of the scale rack and a worker sta-
:o enter the scale after the weighing is completed.

NEG. BN-3149

Figure 6.--A scale platform rack with a side gate for worker who enters and drives animal from the platform to prevent delays.

The sale in the ring is usually conducted by 4 or 5 workers--the auctioneer, recorder, and 2 or 3 ringmen.

Crews ranging in size from 3 to 9 workers yard cattle in buyer pens. However, most markets are arranged so that a crew of 4 or 5 workers can do this work. If an animal is driven out of the sales ring before its sale is completed, it is necessary that one worker temporarily hold the animal in a catch pen adjacent to the sales ring.

Loading-Out Livestock

Although the operations performed in loading-out livestock are similar for all types of markets, markets of type 2 usually load-out cattle at one set of docks and hogs at another. During the sale 1 or 2 workers load-out cattle and 2 or 3 workers load-out hogs. Workers who load out during the sale also receive livestock during this period. Although cattle receipts during the sale usually are comparatively light, hog receipts may be heavy because of the *plus in* system of selling hogs. Under the *plus in* system, the buyer of a particular graded lot of hogs is not only obligated to take all the hogs in the grade at the time of the sale but he also must take all hogs of the grade arriving within a specified time after the sale. After the sale nearly all yardmen usually are assigned loading-out jobs. These crews usually consist of about 14 workers.

Summary of Labor Requirements

The typical market that weighs and sorts hogs on arrival and weighs cattle before their sale (type 2) begins receiving about 7 a.m. and for *graded* hogs continues until about 6 p.m. All 3 cycles of operation are actively conducted during the day. The sale begins about 2:30 p.m. and is completed about 5 p.m. Checking animals for loading out is completed about 7 p.m. Livestock are driven from buyer pens to the loading-out docks throughout the night. The typical market sells animals at a comparatively rapid rate. The total number of workers, other than the office crew, is 18. For the markets studied the number of workers ranged from 16 to 23.

The hog receiving crew is 8 workers. One worker prepares the receiving ticket, 1 worker drives hogs from the dock to the sorting pen, 2 workers sort hogs, 1 worker weighs hogs, and 3 workers yard hogs in buyer pens, all under the supervision of a foreman. When the sale starts 6 workers are shifted to selling operations. Two workers are left in the receiving crew. They also load-out animals during the sale. The cattle receiving crew is 3 workers. One worker writes tickets, 1 worker tags cattle, and 1 worker yards cattle in sellers' pens. When the sale begins 2 workers are shifted to selling operations. One continues to receive cattle. He also loads out during the sale.

The hog selling crew consists of 14 workers: Auctioneer, recorder, 2 ringmen, and 10 yardmen. Of these, 8 workers are obtained from the receiving crews and 6 report for duty when the sale starts. As odd-lot hogs are weighed on arrival, a weighmaster is not needed in this crew. The yard foreman also directs the activities of the yardmen performing selling operations. Two of the 10 yardmen bring up livestock for sale, 3 drive hogs into the sales ring, 1 assigns hogs to buyer pens, and 4 pen hogs in buyer pens. For cattle selling, the hog-selling crew is used, except for the addition of a weighmaster. The weighing of cattle would be performed by the yard foreman. The job assignments of the 10 yardmen are different.

One worker brings up cattle for sale and 4 workers drive the cattle into the sales ring. The number of workers required to assign and pen cattle in buyers' pens is the same as the number required for hogs.

There is no clear-cut division of the labor in checking and loading-out cattle and hogs. The number of workers engaged in loading out each species varies. Workers frequently shift from checking to driving livestock to the truck docks, or 1 worker may perform both jobs consecutively. The loading crew usually consists of 14 workers. After checking is completed and heavy load outs have subsided, 1 or 2 workers may be left on duty throughout the night.

Type 3. Markets that Weigh Livestock Before Their Sale

For the purpose of this study markets that weigh livestock immediately before they are sold are divided into 2 groups because of the wide variation in their other marketing practices. One group of such markets practices dipping cattle, yarding sellers' and buyers' livestock in individual pens, and selling animals both singly and in small lots. Markets in this group are located primarily in Florida. The second group of markets of this type tags and mix-pens 1/ cattle; provides both buyer and seller holding pens; and sells a high proportion of animals singly. Markets in this group are common in north Georgia, North Carolina, South Carolina, and may be found scattered throughout the Southeast. These markets are similar in design to markets that practice after-sale weighing, the primary difference being that the scale is located so that weighing can be done before the animals enter the sales ring.

Figure 7 shows the layout of an auction that practices presale weighing of cattle. A market of this type for which the methods of moving livestock are described would handle about 600 cattle and 50 hogs per sale. Because of the small number of hogs handled, no observations on hogs are included.

Receiving Livestock

Cattle are received by 2 methods. One method involves unloading, separating by ownership when necessary, preparing the receiving ticket, assigning to a holding pen, driving to the dipping vat and dipping and yarding in sellers' pens. Separation by ownership is an intermittent operation, and is usually performed by the worker who assists in unloading cattle. Preparing the receiving ticket usually is done by 1 worker. However, some markets use 2 workers. By this method 22 trucks containing 222 cattle can be received hourly by an 8-man crew, or 2.7 trucks per man-hour.

In the second method dipping is eliminated. By this method, 28 trucks per hour, containing 285 cattle, can be received with a 13-man crew, or 2.1 trucks per man-hour.

These observations indicate that dipping makes but little difference in the number of trucks handled per hour. Receiving crews range in size from 8 to 13 workers.

1/ Mix-penning is the practice of commingling the cattle of several owners in a pen.

- 17 -

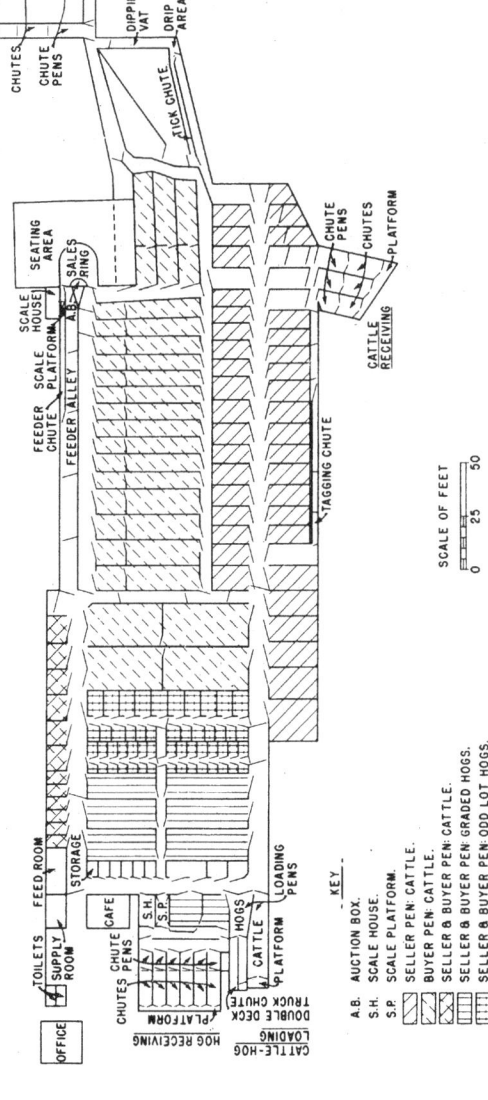

Figure 7.--The layout of an auction that practices presale weighing of cattle. This market also practices weighing and sorting hogs on arrival and dipping cattle.

Selling Livestock

Selling crews range in size from 13 to 23 workers. Pen lots of cattle are brought up from sellers' pens to a catch pen near the sales ring. Because of the large number of nervous cattle, 2 workers normally are used. From the catch pen cattle are driven singly or in small groups onto the scale platform and weighed, and then into a second catch pen adjacent to the sales ring. From this pen they are driven into the sales ring. This crew, including the weighmaster, ranges in size from 5 to 9 workers. To minimize injuries, workers usually operate from walks over the pens to drive cattle onto the scale platform.

Where the scale platform is separated from the sales ring by a catch pen, usually 1 or 2 additional workers are required to drive the animals off the scale platform and into the sales ring.

The ring crew is 4 workers: Auctioneer, recorder, and 2 ringmen. Usually 1 ringman is stationed in the ring, and the other is stationed outside the *out* gate.

One worker assigns cattle to buyers' pens. He calls the pen number of each animal to be yarded. As cattle handled on these markets are not tagged, it is essential that they be yarded accurately. Some markets check livestock as they are being yarded in buyers' pens. Others check while loading out. To check cattle as they were yarded in buyer pens, a worker was located in a position where he could see yarding operations and hear the number of the buyer pen called. This worker acted as overseer of the yarding operation, and saw that errors were corrected. At the conclusion of the sale a chart shows the number of animals yarded in specific pens. When this method was used, no checking was necessary in loading out. The yarding crew ranged in size from 4 to 8 workers.

Loading-Out Livestock

The number of animals in pens of individual buyers first is checked against the number shown on the buyer's invoice. Loading-out crews usually are comprised of 4 workers during the sale. These workers also receive livestock arriving after the sale starts. After the sale, a crew of 12 workers is assigned to loading out.

Summary of Labor Requirements

The typical market that weighs livestock before their sale (type 3) has a yard arrangement about average and sells animals at a comparatively rapid rate. It begins receiving livestock about 7 a.m. and continues until just before the sale ends. The sale begins about 1 p.m. and is completed about 5:30 p.m. Checking animals for loading out is completed by 6:30 p.m. Livestock are driven from the buyers' pens to the loading-out docks throughout the night. The total number of workers, other than the office crew, for the typical market is 23. Other markets of this type employed from 14 to 27 workers. This crew consists of the auctioneer, recorder, weighmaster, 2 ringmen, and 18 yardmen.

The receiving crew is 10 workers. One worker prepares the receiving ticket and 9 workers unload, dip and pen cattle in seller pens, all under the supervision of a foreman. When the sale starts 7 of these workers shift to jobs in the selling cycle. The 3 remaining workers receive and load out during the sale. The number of workers in the selling crew is 19: Auctioneer, recorder, weighmaster, 2 ringmen, and 14 yardmen. Of these workers 7 are from the receiving crew and 12 report for duty at the start of the sale. Two yardmen bring up livestock for sale, 6 drive livestock into the sales ring, 1 assigns livestock to buyer pens, and 5 pen cattle in buyer pens. At the end of the sale 12 workers are shifted to loading-out. These workers usually replace the 3 workers who have been performing these operations during the sale.

DESCRIPTION AND DEFECTS OF MARKET FACILITIES IN THE SOUTHEAST

Some kinds of facilities are used on markets of one type and not on others. A tagging chute is not used if there are individual pens for each consigned lot of cattle. A dipping vat is not provided if cattle are not dipped. When the same kinds of facilities are used, their arrangement varies widely for different types of auctions. The facilities on livestock auction markets in the Southeast can be grouped as follows: (1) Yard facilities, which include truck docks, tagging chutes, facilities for sorting hogs by market grades, holding pens and catch pens, fences and gates, alleys, scales, facilities for driving livestock into the sales ring, catwalks, dipping vats, and testing chutes; and (2) the sales barn, which includes the sales ring, seating area, auctioneer's box, market offices, restaurant or snack bar, and toilet facilities. The yard facilities and sales barn are connected by alleys. Market sites range in size from less than an acre to several acres. In addition, the market site also provides space for driveways and parking areas.

Yard Facilities

Truck Docks

Truck docks, which include platforms, chutes, and chute pens, are the principal facilities used in the Southeast for receiving and loading livestock. The number of docks per auction ranges from 2 to 12. The most common designs are the 2- and 3-level fixed-height dock, and the double-deck loading dock. A 3-level fixed height dock is shown in figure 8.

The 2-level fixed-height dock is designed for pickup trucks and large trucks. The 3-level fixed-height dock provides an additional platform space for semitrailer trucks. Frequently these docks are constructed with only 1 platform space for each type of truck, but in some cases, platform spaces for 2 trucks at 1 or more levels aré provided. The platforms of 2- and 3-level fixed-height truck docks are about 3 feet deep. They usually are constructed of heavy lumber, but in some cases are of rough concrete. Some docks of this type do not have platforms. Chutes for fixed-height truck docks vary in width from about 4 to 12 feet, with 10 feet being most common. Chute fences or sides for fixed-height docks are about 55 inches in height. Gates for fixed-height docks usually are hinged to the side of each chute. They are designed to extend when open across the platform to both sides of the truck tailgate. All dock chutes observed were the ramp type. The floors of the chutes usually are constructed of wood (fig. 9); however, some are of rough concrete. Chute pens always are as wide as and frequently are wider than the chutes. Their depth varied from about 4 to 20 feet. Some docks were constructed without chute pens.

NEG. BN-3150

Figure 8.--A 3-level fixed-height truck dock.

Double-deck docks are confined principally to markets handling relatively large number of hogs. The most common type of double-deck loading dock is of fixed-height. A few docks of variable heights are in use. Dock chutes are about 30 inches wide and the top ramp is 80 to 90 inches above grade.

Improperly designed truck docks frequently result in unnecessary effort by workers to unload and load livestock. On docks without platforms workers must climb on motortrucks to affix chute gates. Where gates do not provide proper enclosures animals occasionally get loose, and time is lost in corraling them. Livestock are more hesitant to move up and down ramp-type chutes than they are step-type chutes (fig. 10). Chutes without pens or with pens of inadequate size necessitate animals being held temporarily on the ramp where they are difficult to manage. Injuries to both animals and workers may occur. Chute pens of inadequate size usually increase labor requirements for sorting livestock by owner lots. When dock heights either are too high or too low for the truckbeds, a slower movement of livestock results.

Trailer Alleys

A trailer alley is a pen usually 12 feet wide and 25 feet deep for unloading livestock from pickup trucks and farm trailers. A truck backs into the alley, the gates are closed, livestock are driven off the truck and into a pen, the gates are opened, and the truck drives out. Trailer alleys are not commonly used in the Southeast, except in a few areas where homemade trailers are popular.

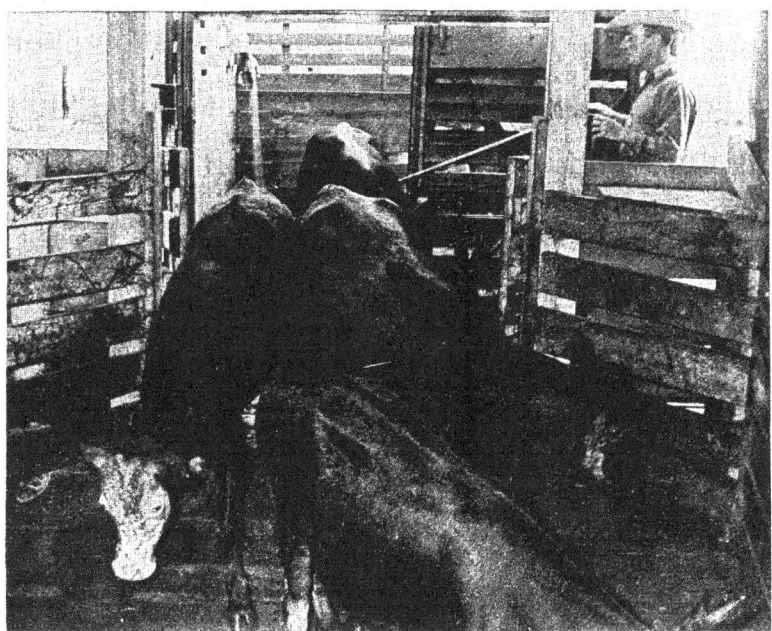

NEG. BN-3151

Figure 9.--Driving cattle down wooden ramp of truck dock.

Tagging Chutes

Special chutes, some *single* and others *double*, are used for tagging cattle at most markets. A *single* chute is usually between 35 and 45 feet long and 30 and 35 inches wide (fig. 11). The sides, usually of heavy lumber, frequently are tapered in width from about 35 inches at the top to 24 inches at the bottom. Gates at each end of the chute restrain animals while tagging is in progress. Floors are either dirt or rough concrete. Taggers either climb onto the side of the chute or work between the upper rails to tag cattle.

A *double* chute consists of 2 parallel chutes separated by a working platform. Their length usually is limited to 20 feet. These chutes are about 36 inches wide and have straight sides. The platform between the 2 chutes is about 24 inches high and about 4 feet wide. Workers operating from the platform tag cattle by leaning over the side of the chute (fig. 12).

The *double* tagging chute has several advantages over the *single* chute. Its shorter length fits better in a market layout and the capacity of each chute is more in line with the size of the lots consigned. The platform between the chutes provides a convenient work station

NEG. BN-3152

Figure 10.--A ramp-type chute on a livestock auction market for unloading livestock.

NEG. BN-3153

Figure 11.--A single-type tagging chute.

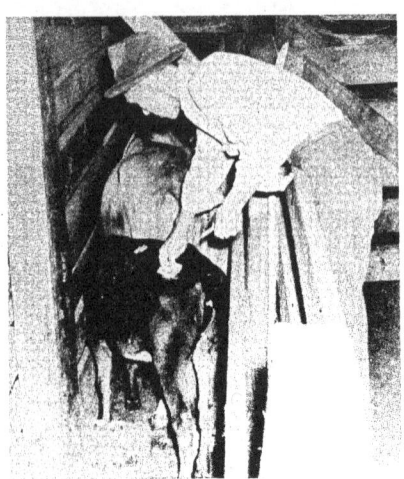

NEG. BN-3154

Figure 12.--Tagging cattle in a double-type tagging chute.

both for tagging cattle and preparing tickets. The double chute also is more conducive to a continuous operation as workers can receive one owner's lot in one side of the double chute while another owner's lot is driven into the other chute.

Facilities for Sorting Hogs

Facilities for sorting hogs into *grades* and weight classes usually consist of a catch pen, a sorting pen, and a scale. Figure 13 shows hogs being driven from the sorting pen onto the scale platform. A few markets have 2 sets of facilities for sorting hogs. Figure 5 shows the layout of a market that has 2 scales and 2 sorting pens. Facilities for sorting hogs vary greatly in design and arrangement and no 2 facilities observed were identical. Hogs are held temporarily in the catchpen, then driven into the sorting pen where they are sorted and driven directly onto the scale platform. The sizes of these pens vary, but usually are at least 10 by 15 feet or larger. The pen fences range in height from 42 to 60 inches. Pen floors usually are dirt, but a few are rough concrete.

Most hog sorting facilities are designed so that animals are driven directly from the sorting pen onto the scale platform.

Most scale platforms are about 7 by 14 feet. Weighbeam enclosures are of 2 types--the *open* type (fig. 14), and the *house* type. The open type is enclosed by a pen about 7 or 8 feet square. This type permits dust, sand, and dirt to settle and collect on the weighbeam and other exposed features of the scale which sometimes results in incorrect weighing and excessive

NEG. BN-3155

Figure 13.--Sorting hogs by market grades.

NEG. BN-3156

Figure 14.--The open-type scale weighbeam setup for hog grading.

maintenance costs. The house type is a house about 8 feet wide, 10 feet long, and 9 feet high. It provides better protection. However, improperly designed scale houses prevent the weighmaster from seeing both ends of the scale platform.

Holding Pens and Catch Pens

Holding pens are used for holding livestock during relatively long periods. Catch pens are used to hold livestock for relatively short periods between drives. Catch pens may be either designated pens in the yards or pens formed in alleys by block gates.

Cattle Holding Pens.--Cattle holding pens usually are laid out in blocks which are separated by an alley the width of which is the same as the pen gate width. Cattle holding pens vary from about 8 to 50 feet in width and from 10 to 80 feet in depth. Pen sizes common to most markets are about 10 by 20 feet, 10 by 30 feet, and 10 by 40 feet. Most auctions have one or more larger pens at the rear or side for handling overflow business on sales days. These pens are as large as 50 by 80 feet.

In the Southeast most cattle pens are under roofs and have dirt floors. However, nearly all markets have some open pens (fig. 15). The floors of these pens are either rough concrete or dirt. Although a few markets have open-trench type drainage systems, most markets do not provide them. Cattle pen sizes generally were too large for auctions to provide individual pens to the large number of sellers requesting them. Large pens also increase the possibilities of injuries to both workers and cattle. In addition they increase the difficulty of handling cattle. Most pens now open on only one alley which may prevent a straight flow of cattle through the pen. The number of holding pens varies considerably. Markets that mix lots may have only a third as many pens as those that provide each consignor and buyer a separate pen.

Hog Holding Pens.--Hog holding pens usually are of 2 types--pens for graded hogs and pens for odd-lot hogs (fig. 5). Pens for graded hogs usually hold 25 to 300 hogs. Pens for odd-lot hogs seldom hold more than 5 or 6 200-pound hogs. Larger pens for graded hogs tend to reduce the amount of handling during receiving and loading-out. However, their use necessitates further sorting in loading as the average number of hogs loaded per double-deck truck is about 70 per deck. Moreover, when large pens are filled to capacity, crowding occurs and injuries may result. Markets that do not sort hogs into market grades usually have only small hog pens.

Some holding pens for graded hogs are about 5 feet wide and 150 feet in length. Others are rectangular or square and range from 10 to 40 feet in width and 10 to 50 feet in depth. The number of these pens per market depends more on the size of the pens than on the volume of business. In 2 markets handling about the same volumes, one may have 25 or 30 pens and the other only 8 or 10. The floors of hog pens are either dirt or rough concrete. Concrete floors are essential for proper sanitation.

As the capacity of hog pens is planned to meet peak sales and as hog marketings are highly seasonal, the utilization of pen space is low. Frequently during summer many pens are empty.

NEG. BN-3157

Figure 15.--Cattle yarded in open pens at an auction in the Southeast.

Catch Pens.--Catch pens for cattle varied in number from 3 to 5 per market. As these pens also are used for hogs, fewer especially designed hog pens are provided. In fact, some markets do not have special hog catch pens. Cattle catch pens usually are about 10 feet wide and 20 or 30 feet deep. Hog pens ranged in size from 3 to 5 feet wide and 5 to 15 feet deep. Most auctions do not have sufficient catch pens to perform efficiently.

Fences and Gates

Cattle Fences and Gates.--Except in areas where large numbers of Brahman and Brahman-cross cattle are sold, cattle fences generally are about 60 inches in height. In areas having nervous and unruly cattle, fences range upward in height to about 7 feet. Cattle fences are constructed of wood with fence rails ranging from 1 by 4 inches to 2 by 10 inches. Rails that are 2 by 6 inches are most common. As many as eight 2- by 6-inch rails are used in constructing a fence 60 inches high but in some cases 9 or 10 rails are used. Often the bottom fence rails are too close to the ground and quickly decay. When too low they also cause an accumulation of manure and dirt. Strong fence construction is common on most markets. Fence construction costs could be reduced an estimated 10 to 15 percent by building fences of proper designs.

Cattle gates are the same height as fences. Because of hard usage, gates often are constructed of heavier materials.

Hog Fences.--Hog fences are usually constructed of wood, but some markets use metal airplane landing mats. Airplane mats, about 42 inches in height, are ideal in height for hog fences (fig. 16). The material is durable and is comparatively easy to apply. Wood fences vary

NEG. IN-3158

Figure 16.--Hog fences of airplane landing mat.

ight from about 40 to 60 inches. Excessive heights are common. Rails usually are 2 by
hes or 2 by 8 inches and frequently are improperly spaced. In many instances 6 rails are
Occasionally fences are constructed with 7 or 8 rails. Both are excessive and increase
ruction costs about 15 percent.

Alleys

Alleys of proper size and location are essential for a free flow of cattle and hogs.

<u>Cattle Alleys</u>.--Most auctions have too few alleys in the cattle sections for the free
of animals. When two or more operations are performed simultaneously in the same alley
s, back drives, and out-of-line drives are frequent.

Alley widths vary from 5 to 14 feet, the most common being 8 or 10 feet. Alleys of
wer widths impede the flow of a full pen of cattle, and animals are bruised. Moreover
w alleys do not permit the movement of trucks for cleaning purposes and hauling feed.
s 12 to 14 feet wide usually necessitate the use of "A" frames or gates of excessive lengths

The number and width of alleys frequently is related to the design of the roof supports.
nimize roof costs market operators often reduce the size and number of alleys, a policy
may lower capital investments but increase operating costs.

Some auctions are laid out with comparatively short alleys which cause undesirable
-angle turns. Cattle often slip when making such turns, causing injuries.

<u>Hog Alleys</u>.--Hog alleys vary in width from 2 to 6 feet, the most common width being 3 or
t. Generally, most alleys are of sufficient width to permit the free flow of full pen
The hog section rarely has more than one main alley and frequently has no cross alleys.
auctions yard hogs after their sale in cattle pens and, at the conclusion of the hog sale,
to stop the sale long enough to drive the hogs back to the pens in which they were yarded
e the sale.

Scales

Most market scales are the weighbeam type with a 5-pound minimum graduation. Where
ed near the auction rings most scales are equipped with a type-registering beam. Scales
ed in connection with sorting facilities do not have these beams. Most weighbeam type
s are equipped with balance indicators. Weighing capacities range from 5,000 to
0 pounds. The scale in most common use has a weighing capacity of 10,000 pounds with a
orm about 7 by 14 feet.

In most cases the weighbeams of scales adjacent to the sales ring are in the auction box
at weight determinations and recordings will be in plain view of both buyers and sellers.
lesign of the auction box or the arrangement of the scale beam sometimes prevents the
master from having a clear view of both ends of the platform. In a few cases the scale
is located in the market office, out of view of buyers and sellers.

Racks built around the scale platforms for cattle or cattle and hogs usually are about
9 feet high, 7 feet wide, and 14 feet deep, with gates at both ends. A gate may be

constructed in one side of the rack. Generally racks are constructed of 2- by 4-inch posts and 1- by 6-inch rails (fig. 17). When used only for hogs scale racks usually are about 42 inches high, 7 feet wide, and 14 feet deep. Some have end gates and some side gates. Racks usually are constructed of 2- by 4-inch or 4- by 4-inch posts and solid sheeted with 1- by 6-inch rails (fig. 18). Both designs require an excessive amount of lumber.

Scale pits are comparatively shallow and repair work is difficult. Figure 19 shows a deep scale pit of excellent design.

NEG. BN-3159

Figure 17.--A scale rack used for cattle and hogs.

Facilities for Driving Livestock into the Sales Ring

All markets make provisions for holding a supply of animals near the sales ring, sorting these animals, and expediting their movement into the ring. The facilities for cattle may differ from those for hogs.

Facilities for Driving Cattle into the Ring.--Although no two markets have identical facilities for this purpose, for cattle they can be broadly grouped as follows: (1) Pens and chute, (2) pens and scale platform, and (3) pens, scale platform, and pen. Pens and chutes are used on markets that weigh animals as they leave the sales ring. Catch pens and the scale platform, or pens on both sides of the scale platform, are used at markets that weigh livestock just before they enter the ring.

The pens and chute arrangement usually consists of a large pen opening into a small pen which in turn opens into a chute. The large pen generally is about 20 by 30 feet, the small

Figure 18.--A scale rack used exclusively for hogs.

Figure 19.--A scale pit easily cleaned and accessible for service or repairs.

pen about 10 by 10 feet, and the chute ranges in length from about 20 to 50 feet. Some chutes are equipped with slide gates at various intervals to prevent cattle from crowding. Chutes usually are about 50 or 60 inches high, 24 to 30 inches wide at the top, and taper to about 18 inches at the bottom. Chutes are constructed of heavy lumber. Cattle are driven first into the large pen, from that pen into the small pen, and then into the chute. From the chute they enter the sales ring. Large lots of cattle to be driven into the large pen for sorting and smaller sorted lots driven into the smaller pen. From this pen they are driven singly through the chute into the ring.

This facility permits an orderly flow of singles into the ring. However, in chutes without slide gates animals frequently crowd causing excessive shrinkage and sometimes severe bruises. Without slide gates animals tend either to move back into the chute when the sales ring door is open, or to rush into the ring 2 or 3 at a time, thus necessitating extra sorting and delaying the sale. When group sales are made more time is required to drive animals singly into the sales ring. Three or 4 men are required to drive animals through the chute. Therefore, labor requirements are excessive. This type facility is rarely used for hogs.

The pens of the pen and scale platform facilities are similar to those in the pen and chute arrangement. In the pens and scale-platform arrangement, the 7- by 14-foot scale platform is substituted for the chute. This facility permits quick handling of both groups and singles, as comparatively large groups of cattle can be driven into the sale ring from the scale platform more quickly than they can from the chute. Hogs also can be handled through this type of facility. By this arrangement the ringman can drive the animals from the scale platform into the ring thus reducing labor requirements. A disadvantage of this facility is the time required to drive single animals from the scale platform into the ring. Scale platforms in use are so large that animals can crowd to one side or to the back when the ring gate is opened and considerable time and effort is necessary to drive them off. On the average, twice as much time is required to drive a single animal from the scale platform as from a well designed chute.

In the third facility described a pen is located between the scale platform and the sales ring. The *out* gate of this pen is the *in* gate of the ring. Pens leading to the scale platform frequently are designed so that workers driving cattle between pens and onto the scale platform can work from the top of rather than in the pens (fig. 20). This facility is found primarily in auctions where large numbers of highly nervous cattle, frequently with horns, are marketed. Its primary purpose is to protect workers from injuries. Pens usually are smaller than the pens in other facilities for driving animals into the sales ring. They range in size from 8 by 10 feet to 10 by 20 feet. Both singles and groups can be driven through this facility. However, the number of workers and the time required are excessive. On an average, one worker drives animals from the scale into the pen and another worker drives them from the pen into the sales ring. Driving the animals from the pen is usually difficult because the pens are so large that an animal can crowd into corners when an effort is made to drive it into the sales ring.

Facilities for Driving Hogs into the Ring.--These facilities consist primarily of 2 or 3 pens, located adjacent to or near the sales ring. These pens insure an adequate supply of animals near the sales ring for continuous selling. They also are used for sorting hogs into salable lots. Pen sizes vary considerably between markets. However, when one pen is from 5 to 8 feet wide and 10 feet deep, the second pen usually is about 5 by 5 feet, or smaller. Large pens are used to handle lots of hogs brought to the sales ring and for sorting the hogs into salable lots. Small pens are used to hold the small salable lots prior to their sale. Auctions that do not sort hogs prior to their sale use these pens to hold lots of hogs.

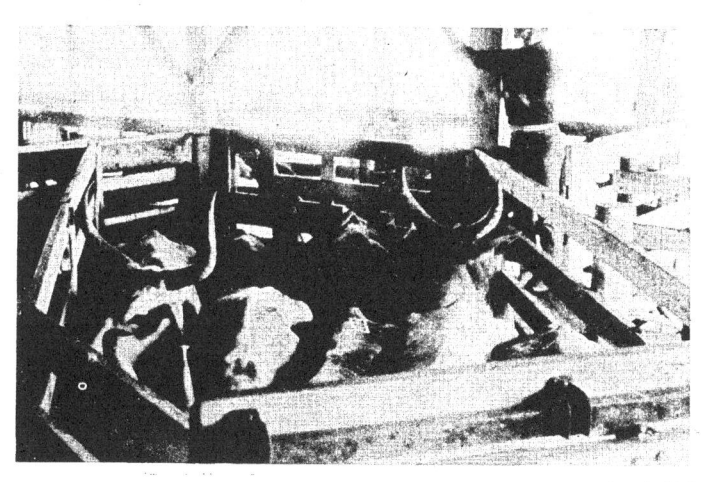

Figure 20.--From an elevated walkway, the worker drives cattle into the sales ring.

Catwalks

Only a few markets in the Southeast have catwalks over the pens to provide buyers, sellers, and spectators an easy, clean, and safe passageway over pens and an unobstructed view of animals in the pens. Without catwalks there are frequent delays when customers and spectators walk through pens and alleys. Moreover customers and spectators are subject to injuries from animals. Catwalks usually are located over the alleys about 4 to 6 feet above the top of the pens. They usually are 36 to 42 inches wide and have a 2- by 4-inch handrail on each side about 42 inches in height (fig. 21).

Dipping Vats

A typical dipping vat is about 72 feet long. The vat proper is about 25½ feet long and 3 feet wide (fig. 22). Its maximum depth is about 6½ feet. The vat is constructed of concrete. Walls above the vat proper are about 7 feet high and constructed of 1-inch lumber. Connected to one end of the vat by a chute 16 feet long is a catch pen about 16 feet deep and 12 feet wide. Connected to the other end of the vat is a drip pen about 14½ feet deep and 10 feet wide. Floors of the drip pen and catch pen also are constructed of concrete.

- 32 -

NEG. BN-3163

Figure 21.--A catwalk above the pens and alleys in a livestock auction market in the Southeast.

NEG. BN-3164

Figure 22.--A vat for dipping cattle.

Testing Chutes

Markets in States requiring that cattle be tested for brucellosis usually provide a holding pen equipped with a testing chute. At one end of the chute a squeeze gate is built to hold cattle while they are tested (fig. 23). Chutes usually are about 28 inches wide and 20 feet long. In some cases a small house is constructed near the chute for veterinarian's supplies and laboratory equipment.

Sales Barns

Some sales barns in the Southeast are pretentious structures designed primarily for market purposes and equipped with air conditioning. Others are comparatively simple structures, some of them having been converted from other uses. Entrances and exits of nearly all sales barns are at ground level, with the main entrances at the front and center of the building. The market office and restaurant usually are at the front. Most sales barns and rings have natural light, however a few are lighted artificially. Usually sales rings are well lighted.

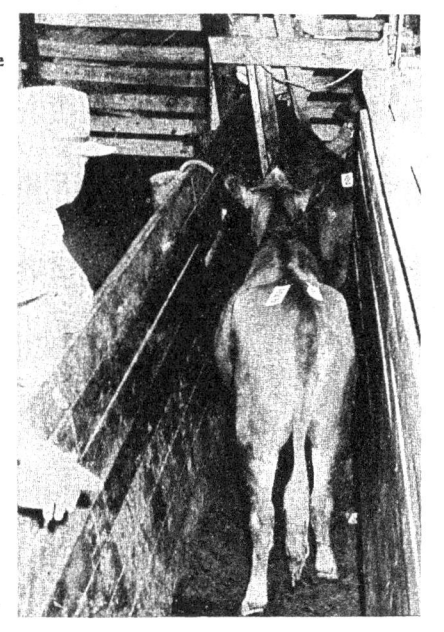

NEG. BN-3165

Sales Rings

Figure 23.--*The animal up ahead is held by the stanchion in the testing chute.*

Many sales rings are "*U*"shaped. Others are rectangular or polygon shaped (fig. 24). Although the dimensions of these rings vary widely, their width (if semicircular) usually ranges from 20 to 32 feet. *In* and *out* gates usually are located on each side of the ring, but in some cases both gates are on the same side. Some sales rings have an 18-inch high concrete wall around the outer edge to provide a base for the construction of the ring enclosure. Enclosures predominantly are constructed of 1- or 2-inch pipe rails 12 to 18 inches apart welded to metal posts spaced at 4- to 6-foot intervals. Most ring enclosures are about 6 feet high, except in areas producing Brahman cattle. In these areas they usually are higher. Some enclosures have a solid wood wall for a height of between 36 or 48 inches with 2 or 3 rails-- either pipe or wood--about 12 inches apart above it.

Most rings cover an area of 300 to 500 square feet. Rings that are too large cause delays in getting animals in and out. The general trend is toward smaller rings.

"*U*"shaped sales rings with *out* and *in* gates on opposite sides usually permit a high degree of efficiency. When these gates are on the same side, an occasional mixup of animals results. More workers usually are required to display and drive animals from rectangular or polygon-shaped rings than from semicircular rings.

NEG. BN-3166

Figure 24.--The sales ring and auctioneer's box in a livestock auction market in the Southeast.

Seating Area

Most market barns provide seats for 200 to 400 people and in some instances the barns are designed also to provide standing room for 150 to 300 people. The seating area is arranged around the sales ring. Usually a small number of seats adjacent to the ring are reserved for larger buyers. When there is only one entrance traffic becomes congested and passage to and from the seating area is difficult. As the turnover of spectators and consignors during a sale is high, congested aisles create a problem. Sometimes the seating area, sales ring, and auctioneer's box are air conditioned, but fans are used more frequently for cooling. Seating areas under low galvanized iron roofs become uncomfortable when temperatures are high.

Auctioneer's Box

The auctioneer's box usually is located across the sales ring directly opposite the seating area between the entrance and exit gates. The counter of the box is about 6 or 7 feet above the floor of the ring. Usually the auctioneer's box has seats for the auctioneer, one or more clerks, and the weighmaster. If the auctioneer cannot see all prospective buyers, he may occasionally miss a bid. A well designed auctioneer's box provides the auctioneer and clerical staff with an unobstructed view of the seating area and the sales ring (fig. 24).

Office Space

Offices usually have space for 4 to 6 employees. A few markets have comfortable and well furnished offices. Others are extremely plain and have but little office equipment. The location of the office is important because of the flow of papers during the sale.

Toilet Facilities

Only a few markets fail to provide toilet facilities in the sales barn.

Market Sites, Roads, and Parking Areas

A major defect of auction markets in the Southeast is the inadequacy of parking space. At some markets insufficient space is available, at others the available space is improperly utilized and unregulated parking results in a confused jumble of cars, trucks, and trailers. It is often difficult for vehicles to reach the loading and unloading docks. Some markets employ workers on sales days to regulate parking. Sometimes vehicles park along highways, thus creating a traffic hazard (fig. 25).

NEG. BN-3167

Figure 25.--Cars of market patrons park along a public road because of inadequate space on the market grounds.

Arrangement of Facilities

The facilities previously described are arranged differently at all auctions in the Southeast. The principal defect of most markets is the poor arrangement of facilities, which results in crossflows of traffic, and necessitates relatively long drives of animals between work stations. A poor layout increases the labor required and retards the rate of sales. A number of markets began in a small way and increased the number of pens and alleys as the volume of livestock receipts increased. Pens usually were added wherever space was available, with little or no consideration of the flow problems that would result, particularly when 3 operations are performed concurrently. If these operations were performed one at a time there would be no particular problem except for long drives at some of the larger markets.

It would be impractical to complete one cycle of operations before beginning another because a substantial increase in the amount of facilities would be required, longer hours and additional man-hours would be necessary, and patrons would become dissatisfied.

The layout of an auction market that weighs and sorts hogs on arrival and weighs cattle before their sale is shown in figure 26. This market sells most cattle singly but occasionally sells small lots.

The total pen area of this market is 23,770 square feet. The hog division has 128 pens containing 13,410 square feet; the cattle division 45 pens covering 10,360 square feet. All pens are under roof. Ten hog pens containing about 5,110 square feet are used for yarding graded hogs, and 92 pens, of about 5,950 square feet, are used for yarding odd-lot hogs. Twenty-six pens, about 2,350 square feet, are used as buyer pens for odd-lot hogs. Eight cattle pens are used as buyer pens, 11 as dual-purpose pens, and the remaining 26 as sellers' pens. The market is equipped with 8 catch pens, 2 sorting pens, 3 scales, a 2- and a 3-level fixed height truck dock, and a double-deck truck dock. The market tags cattle but has no tagging chute.

Receiving Livestock

Figure 26 shows the flow of livestock into the market. Hogs are received at truck docks on one side of the market, cattle at truck docks on the other side. This arrangement is common with markets handling relatively large numbers of hogs and cattle.

Hogs.--After hogs are unloaded, they are driven from the chute pen through alley H 1 to the sorting pen, a relatively short distance. In the sorting pen, hogs are sorted into *grades* and weight classes and each lot driven onto the scale platform. Graded lots are driven through alley H 2 to pens H 1 to H 5 and H 128 and H 129. Pens H 1, H 2, and H 3 are used for yarding hogs grading No. 1, No. 2, and No. 3. When these pens become filled with hogs the entire lots are driven through alleys H 3 and H 4 or H 2 to pens H 6, H 7, or H 8. Yarding graded hogs in pens H 128 and H 129 requires greater effort by workers because the pen gates are not hung for greatest efficiency in yarding through alley H 2. Yarding distances for graded hogs are relatively long and could be shortened by better designed pens.

Odd-lot hogs are driven from the scale platform through alleys H 1, H 5, H 6, H 7, and H 8 to pens H 10 to H 102. The distance to these pens is comparatively long and the gates on pens H 20, H 21, H 37, H 38, H 69, H 70, H 71, and H 72 are not hung for greatest efficiency. When hog sales are exceptionally large, odd-lot hogs and graded hogs are yarded in some of the pens H 105 to H 121, which normally are used as buyer pens. The flow of hogs during receiving is about the same when both sets of sorting facilities are in use.

Cattle.--After cattle are unloaded, those not tagged on the truck are tagged while they are in the chute pen. Animals are driven through alleys 1 and 3 to pens 1 to 18 and 35 to 37. Although these pens are near the truck docks, yarding each individual lot as received requires greater effort by workers than would be required if catch pens were used and cattle driven to sellers' holding pens by pen lots. The gates of pens 1 to 16 and 35 to 37 are not hung for greatest efficiency in receiving. Workers must drive the cattle past the pen, reverse their direction of travel, and then drive them into the pens. Although some lots of cattle are driven from the truck docks to sellers' holding pens 27 to 34 through alleys 1 and 3, the

KEY

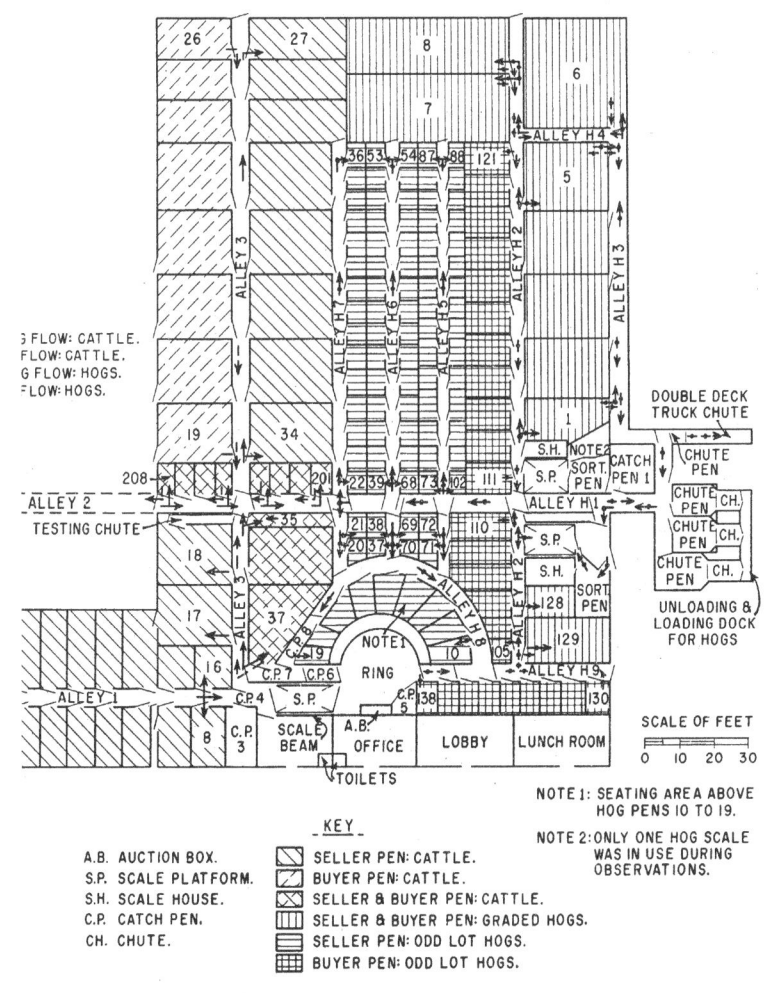

A.B. AUCTION BOX.
S.P. SCALE PLATFORM.
S.H. SCALE HOUSE.
C.P. CATCH PEN.
CH. CHUTE.

▨ SELLER PEN: CATTLE.
▧ BUYER PEN: CATTLE.
⊠ SELLER & BUYER PEN: CATTLE.
▥ SELLER & BUYER PEN: GRADED HOGS.
▤ SELLER PEN: ODD LOT HOGS.
▦ BUYER PEN: ODD LOT HOGS.

NOTE 1: SEATING AREA ABOVE HOG PENS 10 TO 19.
NOTE 2: ONLY ONE HOG SCALE WAS IN USE DURING OBSERVATIONS.

NT OF AGRICULTURE   NEG. 3267-56 (5) AGRICULTURAL MARKETING SERVICE

The flow of cattle and hogs during the receiving and loading cycles at an -ket that practices sorting hogs and presale weighing of cattle.

majority of the animals yarded in these pens are driven from the catch pen through alleys 2 and 3. Cattle yarded in pens 201 to 208 are driven over the same routes.

Selling Livestock

Graded hogs are sold first at this auction, followed by cattle and then odd-lot hogs. Although graded hogs are auctioned in the sales barn, animals are not driven into the sales ring. The market guarantees each grade lot sold. Figure 27 shows the flow of livestock during the sale.

Cattle.--Cattle yarded in holding pens 1 through 16 are driven by pen lots through alley 1 to catch pen 3. Small lots are driven from catch pen 3 to catch pen 4 and from catch pen 4 cattle are driven by salable lots--mostly single animals--onto the scale platform. From the scale platform the cattle are driven in, around, and out of the sales ring through catch pens 6 and 7. The *out* gate to the sales ring is adjacent to the *in* gate of the sales ring. To avoid a mixup of animals in the sales ring, cattle being driven out of the ring must clear the *out* gate before the *in* gate is opened. This arrangement results in occasional delays.

From the sales ring cattle are driven through alley 3 to buyer holding pens 19 to 26. Alley 3 also is the main alley used in driving sellers' cattle to pens 27 to 34. Therefore, during the sale there is interference between sales and receiving operations and delays occur in both operations. When cattle being received become mixed with cattle being sold, sorting is necessary. Pens 201 to 208 and pens 35 to 37 are dual-purpose pens. As alleys 2 and 3 are only 5 feet wide, they are too narrow for driving lots of cattle of the size that can be yarded in the pens opening into these alleys.

Odd-Lot Hogs.--Odd-lot hogs yarded in pens H 20 to H 36 are brought up through alleys H 7 and H 8 to catch pen 8; those yarded in pens H 37 to H 70 are brought up to catch pen 8 through alleys H 6 and H 8, and those yarded in pens H 71 to H 102 are brought up through alleys H 5 and H 8 to catch pen 8. From catch pen 8 hogs are driven into catch pen 7 and, as sale progresses, into catch pen 6. These catch pens are used to maintain an adequate supply of hogs to insure a continuous sale.

From catch pen 6 odd-lot hogs are driven into, around, and out of the sales ring through alley H 9. Buyer hogs are yarded in pens H 130 to H 138, by way of alley H 9, and pens H 105 to H 121, by way of alleys H 9 and H 2. Buyer pens for odd-lot hogs eliminate the need for sorting when loading out. As hog receipts frequently arrive on the market until 6 p.m., their yarding often delays the penning of hogs in buyer pens because alley H 2 is used for both operations.

Loading Livestock

Roughly 20 percent of the cattle and 10 percent of the hogs are loaded out during the sale. Figure 26 shows the flow of livestock out of the market.

Cattle.--Cattle are loaded out at the truck docks at which they are received. Cattle yarded in buyer pens 19 to 26 are driven through alleys 3 and 2 to the loading docks. The gates of these pens hang properly for driving cattle into and down the alley toward the truck docks. However, alley 3 is the major alley for yarding buyers' cattle after their sale; thus,

flow of cattle and hogs during the selling cycle at an auction market that practices sorting hogs and presale weighing of cattle.

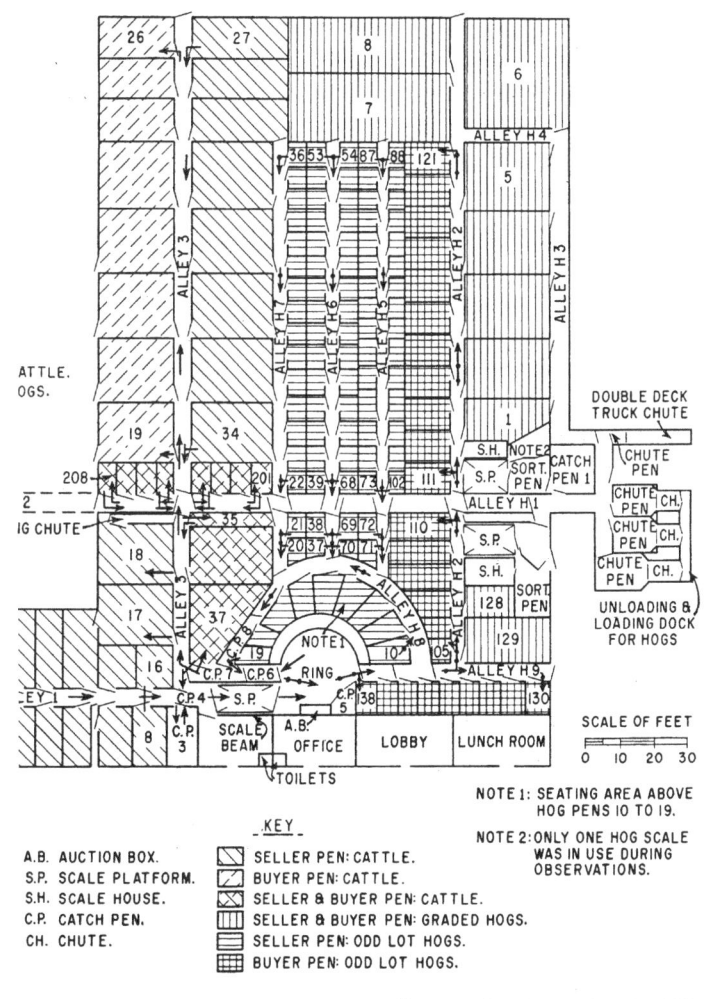

cattle being loaded out from these pens during the sale interfere with the penning-back operation and delay both operations. Cattle yarded in pens 201 to 208 are driven through alley 2 to the truck docks. The gates of these pens are not hung for the greatest efficiency in driving cattle toward the truck dock.

Cattle yarded in buyer pens 35 to 37 are loaded out during the sale by way of alleys 3 and 2. Catch pen 2 serves as a loading pen for cattle. After the sale cattle are driven to the truck docks from buyer pens by way of alley 1, 2, and 3.

Hogs.--Hogs also are loaded out at the same truck docks at which they were received. Graded hogs yarded in pens H 1 to H 5 are driven to the truck docks by way of alley H 3. Graded hogs in pens H 6, H 7, and H 8, are driven to the truck docks through alleys 3 and 4. As the holding capacities of pens H 1 to H 8 are greater than a trailer load or double-deck trailer load, pen lots must be divided into groups of about 70 hogs each a trailer-deck lot. Graded hog pens designed with a capacity for about 70 hogs would eliminate the task of separating the hogs into trailer deck lots.

Odd-lot hogs are driven to the truck docks from pens H 130 to H 138 by way of alleys H 9, H 2, and H 1. Hogs loaded out from these pens during the sale interfere with the yarding of hogs in buyer pens. The same situation occurs when hogs are loaded out from pens H 105 to H 121 by way of alley H 2 and alley H 1.

Summary of Defects

At least 12 major defects in the design and arrangement of livestock auction market facilities affect the efficiency of operation. Although some auctions are modern and operate with a high degree of efficiency, other markets have one or more major defects. The major defects are briefly discussed.

Improperly Designed Truck Docks

Many truck docks either are too high or too low. As a result, livestock often are roughly handled in unloading and loading and movement is slower. Nearly all truck dock chutes are the ramp-type over which livestock move with undue hesitation. Chute pens of inadequate size or a complete lack of pens for holding livestock temporarily at the truck docks are common to a number of auctions.

Inadequate or Improperly Arranged Catch Pens

Some auctions do not have an adequate number of catch pens properly designed and located. Some auctions drive each owner's lot of cattle to sellers' pens on arrival because catch pens are not available to hold these lots until a full pen lot is ready to move. Many catch pens are so located that a reverse flow of livestock is necessary rather than a straight flow into, through, and out of the pen. In many instances catch pens are too large for the small lots which characterize the auction market business. In these pens animals crowd to the side when workers attempt to drive them out. Catch pens that are too large are a waste of space.

Lack of Properly Designed Tagging Chutes

Most tagging chutes are long and narrow permitting the handling of only one owner's lot of cattle at a time. When attempts are made to handle two owners' lots in the chute at the same time, the animals frequently become mixed and sorting is necessary. Tagging and ticket preparation for each owner's lot must be done in sequence in the single tagging chute. This delays the receiving of cattle and results in inefficient utilization of labor.

Holding Pens too Large

Generally, holding pens were designed without consideration of the sizes of lots handled. As a result, many markets cannot grant all the numerous requests of sellers for individual pens. Some market operators report a loss in business because of insufficient number of small holding pens. The crowding of cattle in large holding pens tends to increase the possibilities of losses from injuries and increases the shrinkage losses. Pens too large in size also result in low pen-space utilization.

Improper Type Floors for Holding Pens and Alleys

Dirt floors in a number of markets fail to provide protection against hog diseases. They are difficult to clean, to keep free of odors, and to drain.

Excessive Fence and Gate Construction

A number of auctions have fences in the hog section as high as fences in the cattle section. This added height is of no practical value in handling hogs. Many of the fences and gates in both the hog and cattle sections are constructed with more rails than are needed. Excessive construction causes greater outlays of investment capital and generally results in higher maintenance costs.

Improperly Located Scales

Too frequently the location of scales necessitates the use of an unusually large number of workers for driving livestock on and off the scale platform. Weighbeams of many scales are located in the offices of the auctions or to the rear or side of the auctioneer's box where buyers and sellers are unable to see the weight determinations and recordings.

Lack of Properly Designed Facilities for Driving Livestock into the Sales Ring

Facilities used for driving cattle and hogs into the sales ring frequently fail to maintain the animals in proper position for entering the sales ring quickly and to prevent them from crowding. As a result, sales are delayed and the possibilities of injuries are increased.

Improperly Arranged Yard Facilities for a Direct Flow of Livestock Through Market Operations

Improperly arranged yard facilities cause: (1) Long drives of livestock between work stations, (2) crossflows, (3) mixups of livestock, and (4) back drives and out-of-line drives of animals.

Improperly Designed Sales Ring

Sales rings generally are too large for handling small lots which characterize auction sales in the Southeast. Consequently, ringmen are unable to drive animals from the rings promptly. Occasionally additional workers are added to the ring crew to assure a speedy sale. In square shaped sales rings animals tend to crowd into the corners when workers attempt to drive them from the ring.

Poorly Designed Sales Barns

The sales barns on some auctions originally were constructed for other purposes. In many instances the seating arrangement is inferior and the seats uncomfortable. Prospective buyers cannot obtain a clear view of the animals in the ring and the auctioneer is unable to see prospective buyers. Some barns do not permit free passageway through them during the sale; entrances and exits are congested with spectators, buyers, and sellers.

Small Market Sites

Many markets, particularly some of those that have been established for several years, have sites of inadequate size and, as a result, traffic is congested, delays occur in unloading and loading of motortrucks, and insufficient parking space is available.

SUGGESTED DESIGNS FOR LIVESTOCK MARKET FACILITIES IN THE SOUTHEAST

Some of the facility designs suggested in this section are based on facilities now in use in the Southeast. Some have been redesigned to correct obvious defects. Other suggested designs are based on research on both auction markets and terminal stockyards. Although a number of the facilities described are used by all types of auction markets, some are used by only one type. Nearly all the facilities suggested, however, should have widespread application in all areas of the United States.

Truck Docks

Livestock are received and loaded out on pickup trucks, farm stake-body trucks, and tractor-trailer trucks, all of which vary in height of floor or bed. Trucks docks of varying heights and widths are needed. Most markets need at least 1 truck dock space for each of the 3 types of trucks. Two types of docks appear best to meet the need on markets in the Southeast: The 3-level fixed-height dock and the double-deck dock.

ure 28 shows the suggested design of a 3-level fixed-height dock. This facil
bridge differences in the heights of floors of vehicles and floor of the mar

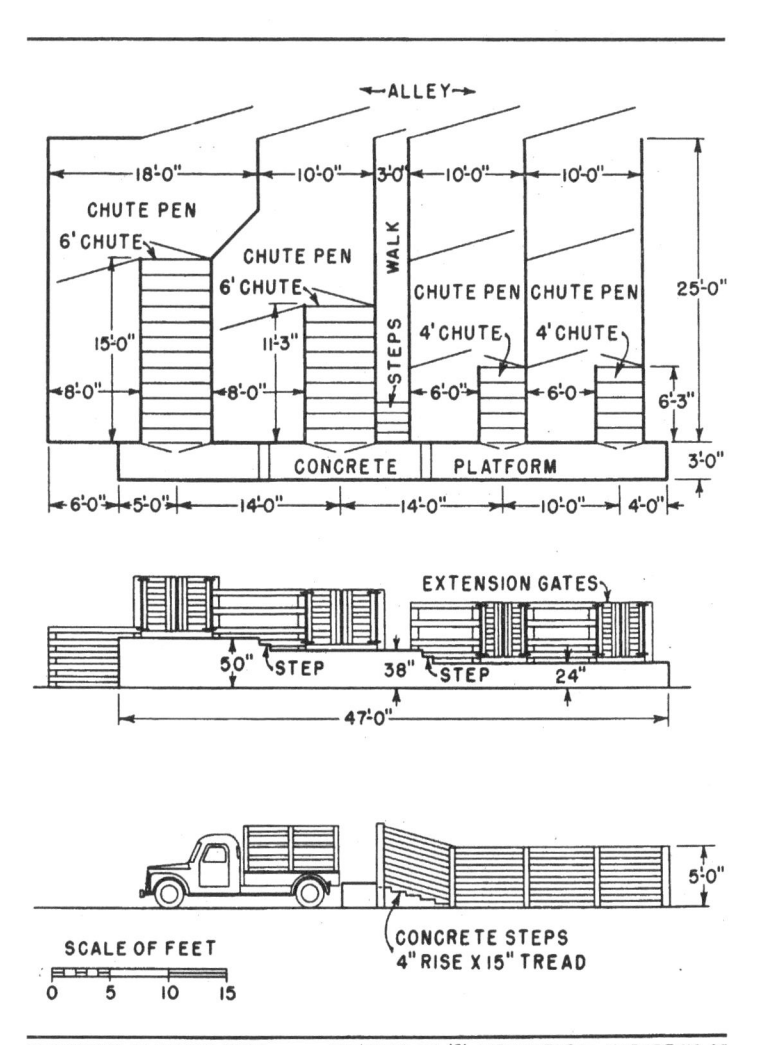

Figure 28.--Suggested design for a 3-level fixed-height truck dock.

This design includes 2 *spaces* for pickup trucks and one each for farm stake-body type trucks and tractor-trailer trucks. Its overall dimensions are 47 feet wide and 28 feet deep. The dock has a continuous front platform 3 feet deep which enables workers to unload trucks easily and quickly. The platform also provides a convenient work station for the preparation of the receiving ticket and for tagging cattle. Step-type chutes are suggested because livestock move up and down steps with less hesitation than over cleated ramps. Chute fences and gates should be similar to those shown later in this section for various types of pens. Platforms should be constructed of reinforced concrete. (For a more detailed discussion of truck docks, see Agriculture Handbook No. 36, *Suggestions for Improving Services and Facilities at Public Terminal Stockyards,* U. S. Department of Agriculture, January 1952.)

The section of the platform for pickup trucks should be 24 inches high. Each of the 2 chutes connecting this part of the platform and the chute pens should be 4 feet wide and 6 feet 3 inches deep. The platform width from center to center of the 2 chutes should be 10 feet. An allowance of 4 feet at each side would provide a total of 18 feet and permit 2 trucks to unload or load simultaneously. Both chutes should lead directly into chute pens. Double extension gates should be provided at the platform end of each chute and full-width gates at the chute pen end. As the total pen area of 225 square feet suggested for each chute is too large for the small loads received on pickup trucks, it should be divided into 3 chute pens--2 containing 94 square feet each and one containing 37 square feet.

The center section of the platform suggested for farm stake-body type trucks should be 38 inches high. A single step having a 7-inch rise should connect this section with the 24-inch-high section. The single chute connecting the platform and the chute pen should be 6 feet wide and 11 feet 3 inches deep. Chute steps should be constructed with 4-inch rises and 15-inch treads. The chute area containing 250 square feet should be divided into 2 pens, one of 160 and the other of 90 square feet. Three-foot double gates each with 18-inch extensions should be provided at the platform end of the chute and a single 6-foot gate at the pen end. The distance from the center of this chute to the centers of those on each side should be 14 feet.

The section of the platform suggested for tractor-trailer trucks should be 50 inches high. The single chute should be 6 feet wide and 15 feet deep. Chute steps should be constructed with 4-inch risers and 15-inch treads. The chute pen area containing 292 square feet should be divided into 2 pens, one of 172 square feet and one of 120 square feet. Three-foot double gates with 18-inch extensions should be provided at the platform end of the chute and a single 6-foot gate at the pen end. The distance from the center of this chute to center of next inside chute should be 14 feet.

If the outside truck dock approaches can be depressed so that the dock platform will be level with the floor of the holding pens and alleys and with the beds of all 3 types of trucks, no chutes are necessary. Animals would be unloaded directly across the platform into the chute pens. This type of dock is not suitable for many markets because of drainage problems. Figure 29 shows a suggested design for truck docks where the truck approaches can be depressed. Except for the elimination of the chutes, this dock is similar to the one shown in figure 28.

The suggested depth of this dock is 18 feet. Two chute pens, each 10 by 15 feet, are suggested for pickup trucks. Pen gates opening on the platform should be 2 feet wide and equipped with 24-inch extensions. The *out* gates opening into an alley should be 10 feet wide. The distance from center to center of the 2 double gates on the platform side should be 10 feet.

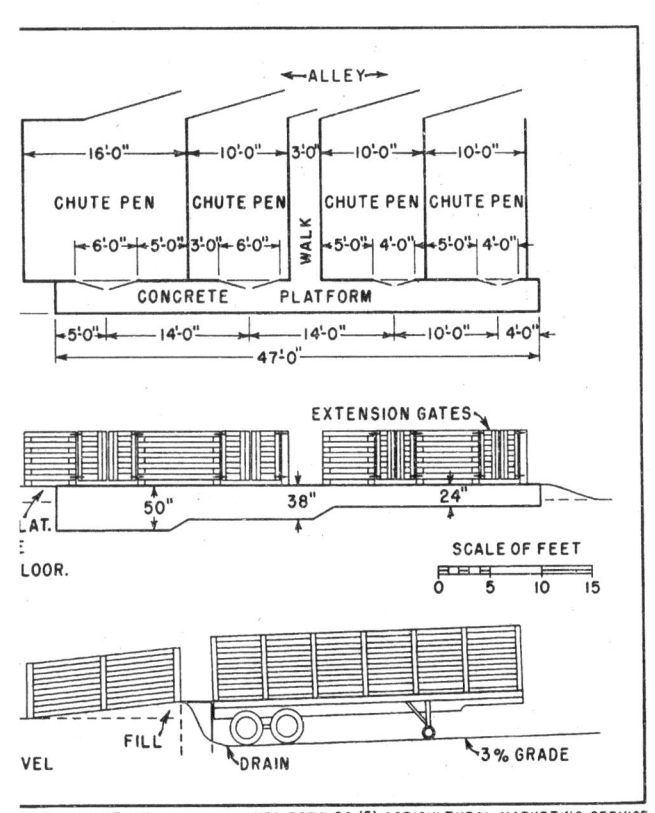

gested design for a truck dock where approaches can be depressed to
provide proper truckbed height.

on the platform side of the section for farm stake-body type trucks and
would be 3 feet wide and equipped with two 36-inch extension gates. The
section also should contain 150 square feet. The distance from center to
pen gates should be 14 feet. The section of the dock for tractor-trailer
comparable to the center section except that the chute pen should be 15 by

trucks require considerable space for maneuvering within the dock area,
should be about 150 feet deep and 100 feet wide. The depressed area

should not exceed a 3-percent slope and should be well drained and surfaced with gravel or other materials to stand up under the weight of trucks.

The double-deck truck dock (fig. 30) is used primarily for loading-out hogs to double-deck trucks. The suggested design consists of 2 stepped chutes, built side by side, which lead from loading pens to the upper or lower decks of a truck. The dock platform is 8 feet wide, 3 feet deep, and 50 inches high. The overall depth of this dock is about 30 feet. Both decks of a truck can be loaded at the same time.

Each chute is 3 feet wide and constructed of steps having 3-inch risers and 10-inch treads. The upper deck chute is 23 feet, 4 inches deep in order to reach a height of 86 inches above the ground, the average height of the upper decks of trailer trucks. The lower deck chute rises 50 inches or to the level of the platform. The upper deck chute has a hinged ramp 3 feet wide and 3 feet deep, and 2 wing gates--each 3 feet wide with 2-foot slide extensions. When lowered the ramp rests on the upper deck of the truck. The lower deck chute has 2 wing gates, each 5 feet wide with 3-foot slide extensions.

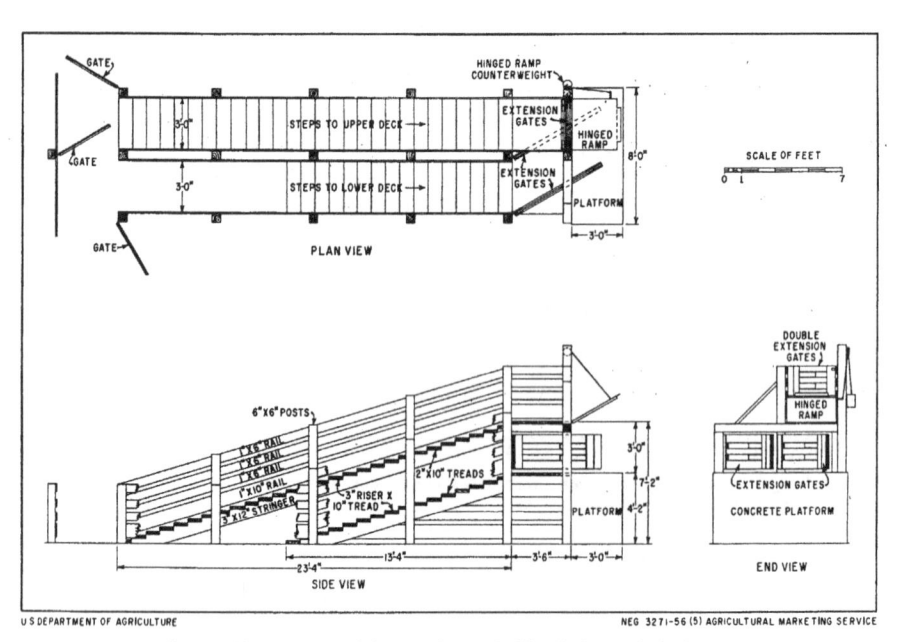

Figure 30.--*Suggested design for a double-deck truck dock.*

Cattle Tagging Chute

njuries to workers and relatively high labor requirements,
y trucks and tractor-trailer trucks should not be tagged
safely and minimize labor requirements, a special chute o
should be provided.

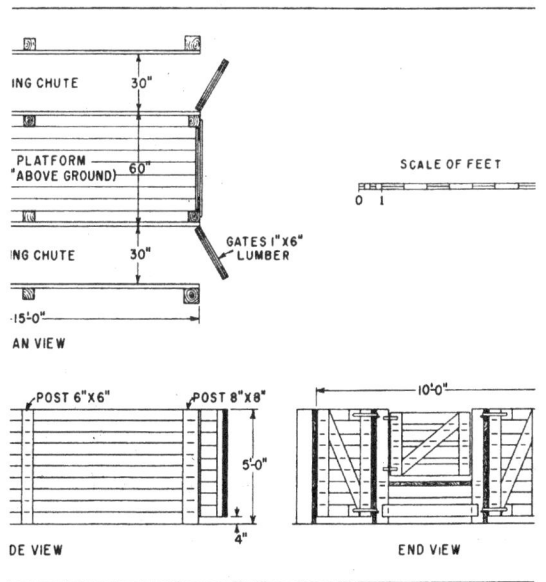

1.---Suggested design for a cattle tagging chute.

narrow alley through which animals can be moved in single
rised of 2 chutes, each 30 inches wide at both top and bot
ong, separated by a work platform 5 feet wide and 22 inch
are 30-inch gates. Solid boarded sides keep the horns of
n rails. Cattle are driven in one end of the chute, tagged
d. Each chute holds about four 500-pound cattle.

more continuous performance since tagging and ticket prep
wner's cattle in one chute while another owner's cattle a
te.

Cattle Fences and Gates

Cattle fences at auctions where nervous cattle are predominant should be 7 feet high. At other markets they should be 5 feet high. Line posts should be either 6-inch top round or 6- by 6-inch top squared material spaced not more than 7 feet apart. Corner and gateposts should be either 8-inch top round or 8- by 8-inch top squared material. All pine posts should be pressure treated with either coal-tar creosote or a petroleum solution of pentachloraphenol. Fence rails should consist of either 5 pieces of 2- by 6-inch lumber or 4 pieces of 2- by 8-inch lumber. The lower rails should be 8 inches above ground level. Two hip rails of 2- by 8-inch material on the inside of the fence will minimize losses from bruises. The lower hip rail should be 28 inches above the ground. Figure 32 shows the suggested design and spacing of rails for a 5-foot cattle fence. In areas where nervous cattle are not too prevalent, consideration might be given to the use of full sawn 1- by 6-inch boards for fence rails.

Gates for cattle pens should be strong and durable. Actually, many gates--particularly the gates of buyers' holding pens--are opened as many as 30 times on sales days. In extreme cases certain gates are opened as many as 60 times. Cattle gates should be 5 feet high except on auctions where nervous cattle predominate. Rails for cattle gates should consist of 4 pieces of 2- by 8-inch lumber and the lower rail should not be less than 8 inches above the ground.

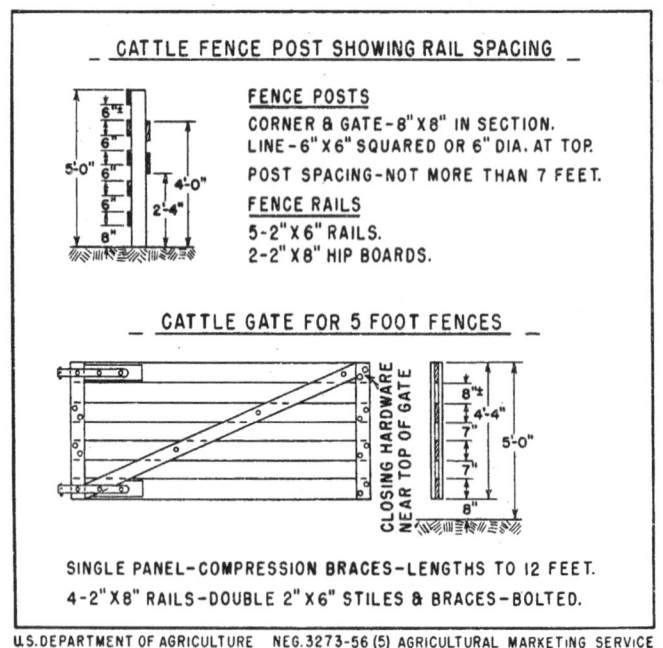

Figure 32.--Suggested design for a 5-foot cattle fence and gate.

Hog Fences and Gates

s for hog pens should be 3½ feet high. Line posts should be 4-inch top round or
ch top square material, set not more than 5 feet 3 inches apart. Corner and gate.
ld be 6-inch top round or 6- by 6-inch top square material. All pine posts should
with either creosote or pentachloraphenol. Rails should be placed on each side of
. Fence rails should consist of one 1- by 10-inch piece and three 1- by 6-inch
lumber for each side. The lower rails should not be less than 4 inches above ground
In some instances steel landing strip material is used to construct hog fences.
ial, which is manufactured in strips about 6 feet long and 3½ feet wide, is made up
ced strands of 3/4-inch strip steel crossed at 4-inch intervals.

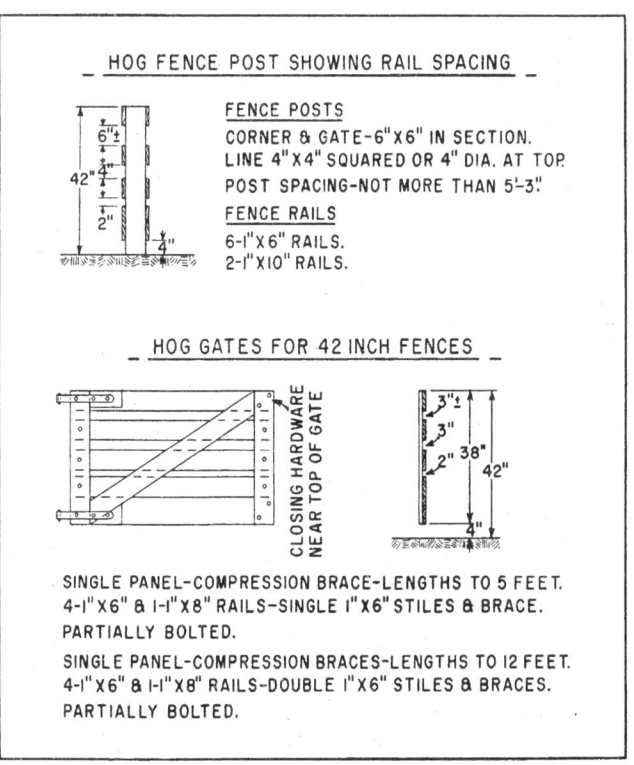

HOG FENCE POST SHOWING RAIL SPACING

FENCE POSTS
CORNER & GATE-6"X6" IN SECTION.
LINE 4"X4" SQUARED OR 4" DIA. AT TOP.
POST SPACING-NOT MORE THAN 5'-3".
FENCE RAILS
6-1"X6" RAILS.
2-1"X10" RAILS.

HOG GATES FOR 42 INCH FENCES

CLOSING HARDWARE NEAR TOP OF GATE

SINGLE PANEL-COMPRESSION BRACE-LENGTHS TO 5 FEET.
4-1"X6" & 1-1"X8" RAILS-SINGLE 1"X6" STILES & BRACE.
PARTIALLY BOLTED.

SINGLE PANEL-COMPRESSION BRACES-LENGTHS TO 12 FEET.
4-1"X6" & 1-1"X8" RAILS-DOUBLE 1"X6" STILES & BRACES.
PARTIALLY BOLTED.

U.S. DEPARTMENT OF AGRICULTURE NEG. 3274-56(5) AGRICULTURAL MARKETING SERVICE

Figure 33.--Suggested design for a 42-inch hog fence and gate.

Gates for hog pens also must be of durable construction. They should be the same height as the fence. Gates should be long enough to close the alleys into which they open. Rails for hog gates should consist of 4 pieces of 1- by 6-inch lumber and one piece of 1-inch by 8-inch lumber; lower rail should be about 4 inches above the ground. Gates should be constructed with compression braces.

Scale Platform and Rack

A suggested design for a scale platform and a platform rack for cattle is shown in figure 34. The scale platform is 7 by 14 feet. It should be constructed either of reinforced concrete with a rough surface, or of 2-inch lumber on weighbridge girders and covered with a layer of waterproof roofing paper and a 1-inch wearing surface. Two- by 2-inch wooden cleats, spaced 12 inches apart, should be fastened on the wearing surface to make it nonskid.

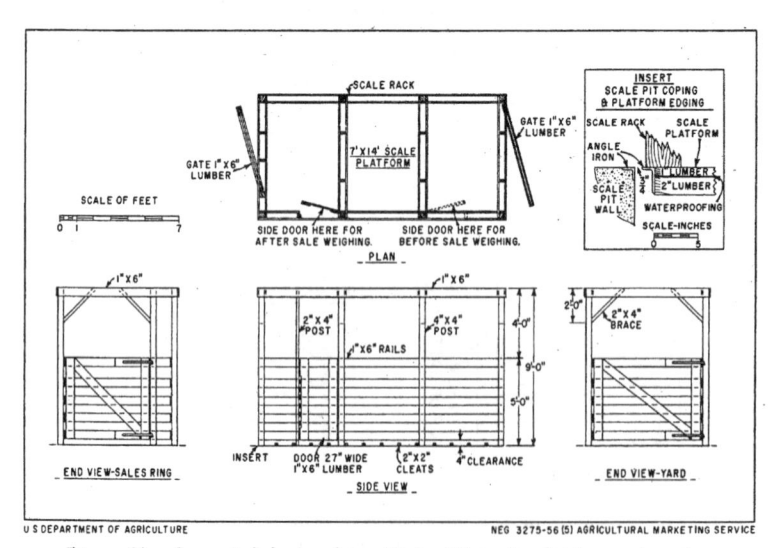

Figure 34.--Suggested design for a 7' by 14' scale platform and rack.

The scale rack for cattle should be constructed of 4- by 4-inch posts, 9 feet high and spaced at equal distance along each side, and 5-foot high walls constructed of 5 pieces of 1- by 6-inch lumber. The lower rail should be 4 inches above the platform floor to prevent the accumulation of debris and permit easy cleaning. Gates of the same height as the sides should be constructed at each end of the rack. The gates should be full width. To permit a worker to drive livestock from the scale platform quickly, a 27-inch gate should be placed in the side of the rack.

re the rack rigidity, posts should be braced at the top with cross, knee, and
braces. These braces should be of 1- by 6-inch and 2- by 4-inch lumber.

ggested design of a scale platform and rack for hogs is shown in figure 35. The
by 4-inch posts spaced at equal distance along each side of the platform. Door
oor stops also are of 2- by 4-inch lumber. The sides of the rack should be 42 inche
nstructed of seven 1- by 6-inch rails. On the sorting pen side of the rack there
wo 4-foot gates hung to a common 4- by 6-inch post in the center of the rack side.
site side of the rack there should be two 5-foot wood gates hung about 42 inches
ng in opposite directions. To make the scale rack rigid the rack posts should be
et high with braces at the top. Braces should be of 1- by 6-inch and 2- by 4-inch

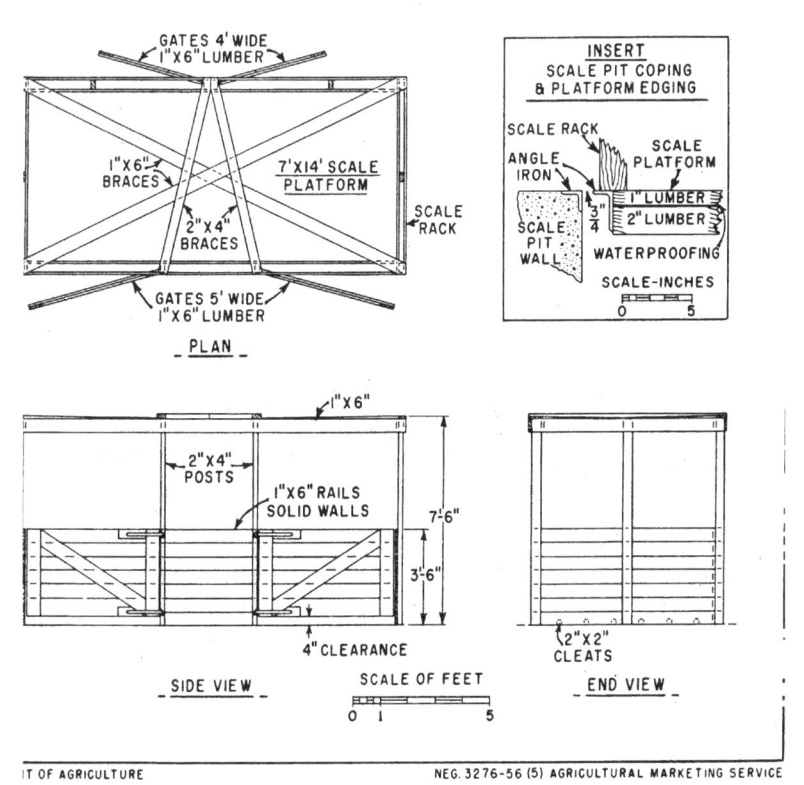

re 35.--*Suggested design of a scale platform and rack for sorting hogs.*

- 52 -

Pits for scale platforms and racks should be 6 feet deep, 7 feet 1½ inches wide, and 14 feet 1½ inches long. They should be constructed of waterproof concrete. Although no drainage is required, pits should be equipped with a sump pump. The pit should be constructed so that repair can be done while the scale is in operation. It should have an electrical outlet.

Feeder-Chute Facility

A feeder-chute facility is used to maintain an adequate supply of animals near the sales ring and to permit both singles and groups of animals to be driven into the sales ring quickly and in an orderly way. One man inside the feeder chute can control several cattle and drive each animal into the ring as quickly as it is needed.

Figure 36 shows the suggested design of a feeder-chute facility, which is 25 feet long and 9 feet wide (inside dimensions). It is divided into 3 lanes--a 5-foot wide center lane which is paralleled on each side by 22-inch (inside dimension) outer lanes. Double gates--each roughly

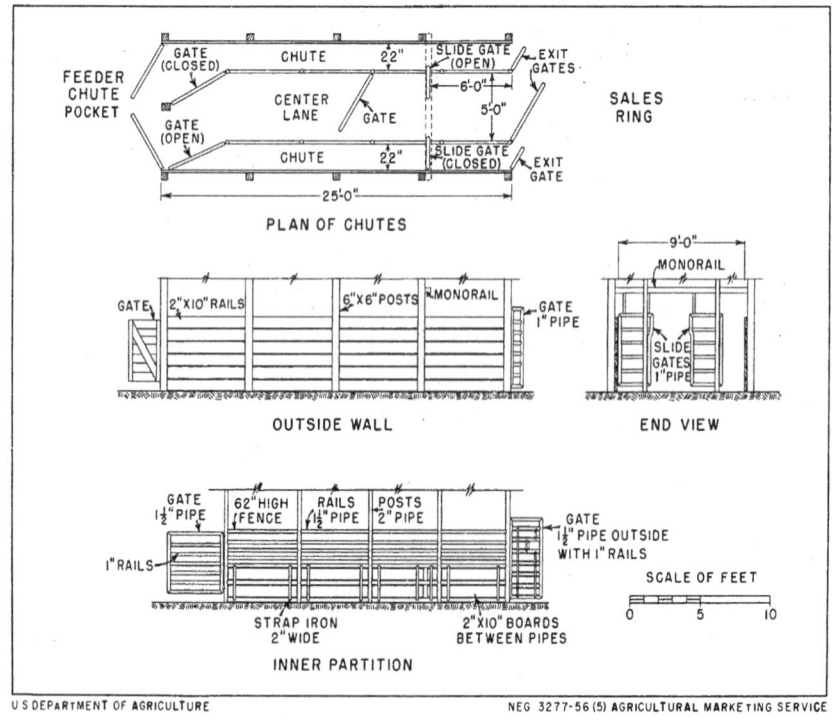

Figure 36.---Suggested design of a feeder chute for driving animals into sales ring.

4½ feet wide--are used at the entrance of the chute leading from the feeder-chute pocket. These gates are hinged toward the outer chute partitions and close against a post stop in the center of the chute. Double gates of approximately the same width are set back about 4 feet from the entrance and are swung from the inner partitions to close against the inner side of the same post stop when shut and against the outer partitions when open. Opening one outer gate and closing the inner gate diverts animals into an outer lane. Opening both sets of gates diverts animals into the center lane. At the exit ends of the chutes, single, hinged gates are used for each lane. Slideblock gates in each of the outer lanes 6 feet back of the exit gates operate on an overhead monorail. Temporary gates also may be placed in a comparable position in the center lane for use with hog sales.

The outside walls of the chute should be constructed of 6- by 6-inch posts spaced 6 feet apart on center and solid faced on the inside from the ground level up with 6 pieces of 2- by 10-inch lumber. The inner partitions of 2-inch iron pipe posts are spaced 5 feet apart, and at intervals 1½-inch pipe are welded to them to form rails. Two pieces of 2- by 10-inch lumber are strapped inside of and flush with the surfaces of the lower rails.

The outer lanes of the chute are used when singles are sold. Each lane will hold 5 to 7 animals. The slide gates in the outer lanes are used primarily to permit single animals to enter the sales ring instantly and to assist in preventing animals from crowding in the chute. The center lane is used when groups of animals, bulls, and large single animals are sold. The center lane also is used for hogs. When closed, the gate in the center lane forms 2 pens. Hogs can be sorted into salable lots in one pen and then driven into the other pen to be temporarily held.

One worker in the feeder-chute pocket drives animals into the chute, and one worker inside the chute drives them into the sales ring. In addition to its low labor requirements, this facility offers maximum protection to the workers. Moreover, it requires less space than some other facilities.

Cattle Trough

Figure 37 shows the suggested design of a reinforced concrete water trough for cattle pens. This trapezoidal trough is 10 feet long, 12 inches high, 14 inches wide at the bottom, and 22 inches wide at the top. A constant level drain pipe can be removed for cleaning the inside of the trough.

Hog Trough

Figure 38 shows the suggested design of a reinforced concrete water trough to be located parallel to and under a fence separating 2 adjacent hog pens. The rectangular trough is 20 inches wide and 8 inches high. Its length is determined by the length of the pens. The constant level drainpipe can be removed for cleaning purposes. The trough should be deeper at the drain end than at the other end.

Hayrack and Grain Trough

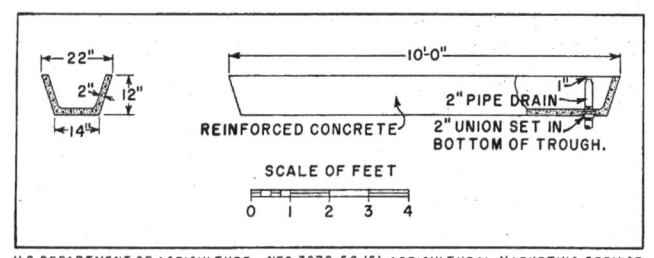

Figure 37.--Suggested design of a water trough for cattle.

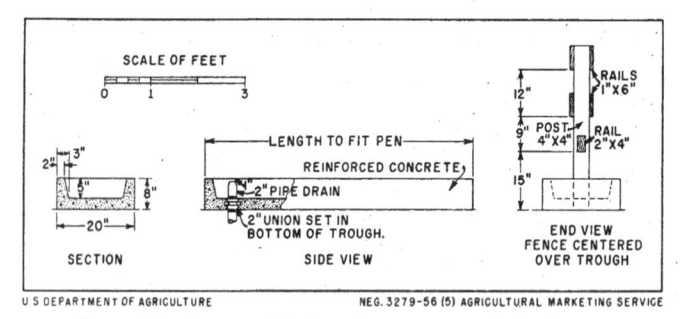

Figure 38.--Suggested design of a water trough for hogs.

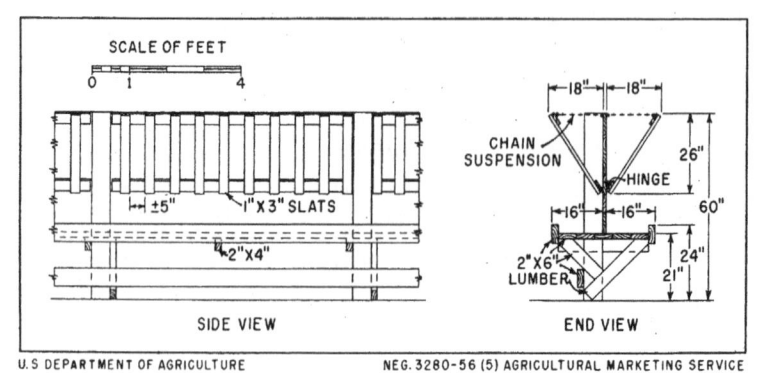

Figure 39.--Suggested design of a double, combination hayrack and grain trough.

ugh floor should be 21 inches above ground level. Each trough should
nches deep. Hayrack slats should be 30 inches long, 3 inches wide, a
rt. The suggested length of the hayrack and grain trough is 20 to 25

Catwalks

Catwalks provide buyers, sellers, spectators, and others a means o:
hout walking through alleys and climbing fences. They should be cons1
 ground level and preferably afford an unobstructed view of all pens
ks should be supported on 6- by 6-inch or 8- by 8-inch posts, across
ced 2- by 6-inch stringers (fig. 40). Catwalks should form a complet(
ds. Stairs to connect the catwalks with the ground level should be p]
 auction barn near the truck docks. Catwalks should be 4 feet wide a1
dom-width boards supported above the stringers by three 2-inch by 6-i
eet high, and constructed of 2- by 4-inch lumber, should be provided
kway.

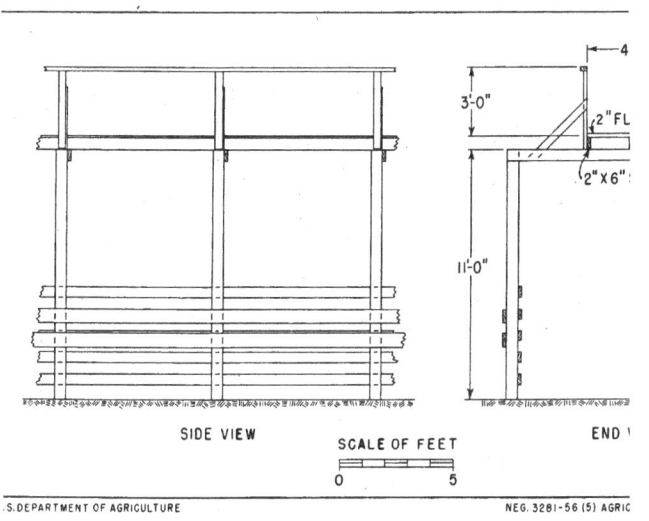

Figure 40.--Suggested design of a catwalk over market yo

Auction Barn

The auction barn and the operations performed in it are appraised b
ι any other part of a market. Therefore, barns should permit operatin
·ide conveniences for buyers, sellers, and spectators, in line with m

business practices. Principles to observe are: (1) Design barns requiring minimum construction and maintenance costs. Most barns are used for sales only 1 day each week. Their use for other purposes is limited. (2) Design barns that provide for a free flow of pedestrian traffic into, through, and out of the facility and meet the requirements of local ordinances. As buyers, sellers, and spectators constantly go into and out of the barn during a sale adequate entrances and exits are essential. (3) Provide accommodations in barns for an audience of fluctuating size. As the number of people attending sales fluctuates, it is impractical to provide seats for peak audiences. It is more practical to provide some seats and a balcony where people can stand and view the sales ring. (4) Arrange seats so buyers, sellers, and spectators have an unobstructed view of the sales ring and auctioneer's box. (5) Design a sales ring that permits a free flow of livestock into, through, and out of it with a minimum of labor. (6) Locate market offices for easy accessibility to buyers and sellers and to facilitate the exchange of records with the auctioneer's box.

To provide adequate floor area for the components of the auction barn, the barn proper should be 66 feet wide and 45 feet deep. At the rear of the building an attached covered annex 9 by 10 feet should provide space for the auctioneer's box.

As shown in figure 41 the main entrance to the suggested sales barn leads into an 8- by 12-foot lobby on the first floor. To the right and left of the main entrances are stairways leading to the sales-barn auditorium on the second floor. The overall height of the building is 20 feet at the eaves.

The offices, restaurant, and toilets are at ground level under the balcony and seating area. Six entrances to the seating area are provided: Two by the stairs at the front of the building, 1 on each side, and 2 from catwalks at the rear. The 2 side entrances are at ground level. Entrances at the front open into the balcony at the second-floor level. It is suggested that the barn be of frame construction.

Sales Ring

The sales ring should be of proper size and design so that: (1) Buyers may be given ample opportunity to see the animals offered for sale; (2) the desired rate of selling may be maintained; (3) the efficiency in performing the operation can be increased; and (4) both market employees and livestock will be protected from injuries. The ring should be large enough to accommodate a large truckload of cattle without crowding the animals and small enough to handle singles efficiently. The ring should be of a size that physical handling operations in connection with sales can be performed efficiently by 2 workers. The suggested sales ring is semicircular. The base of the semicircle should be 23½ feet and its perimeter at the highest point from the center should be 9½ feet. Safety islands are provided on each side of the ring. Each island should be 18 inches from the ring fence and should consist of 2 posts 6 inches in diameter and 5 feet high, constructed on 16-inch centers. Entrance and exit gates, separated by the auctioneer's box, should be placed at the outer edges of the base. The entrance gates to the ring are the 3 exit gates from the feeder chute previously described. These gates occupy roughly 9 feet of the 23½ foot base. A single 5-foot exit gate leads onto the scale platform. The counter of the auctioneer's box projects 3 feet into the ring, leaving a net depth of 9½ feet inside the ring.

The fence separating the sales ring from the seating area should be 7 feet high; for Brahman cattle, it should be at least 8 feet high. Posts for the sales ring fence should be of

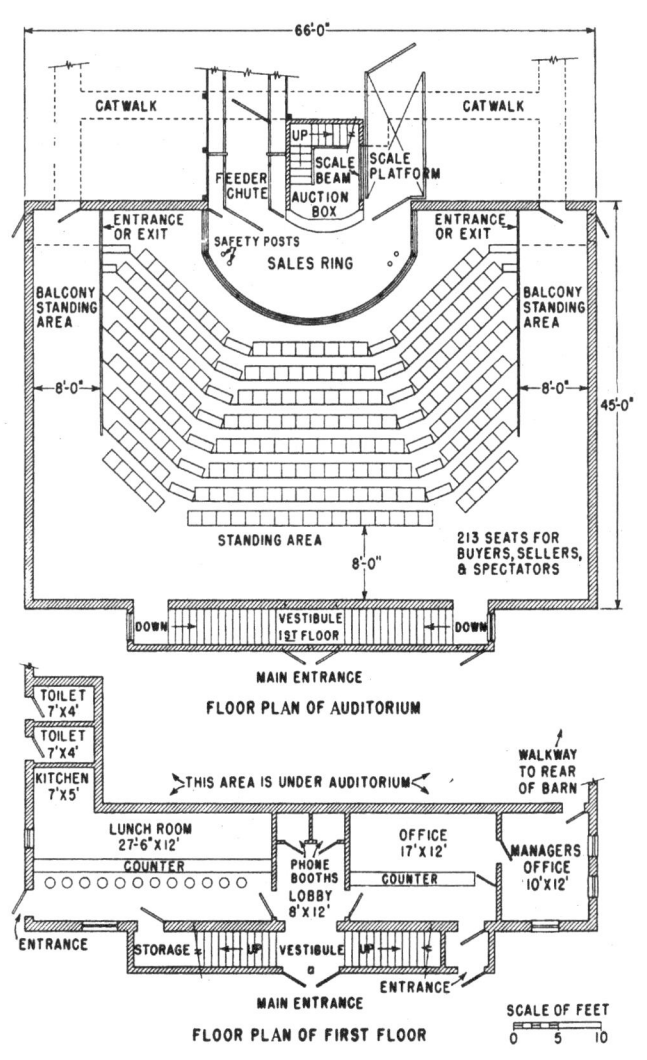

Figure 41.—Floor plan of the suggested sales barn for livestock auction markets in the Southeast.

2-inch galvanized steel pipe spaced at 6-foot intervals and embedded in an 8- by 24-inch concrete base. Fence rails should be of steel pipe, varying in diameter from 1½ inches for the lower rails to 1 inch for the upper rails, welded at right angles to the post to form horizontal bars. At the base of the fence, the rails should be spaced 6 inches apart and the spaces between bars progressively widened until the top bars are 12 inches apart.

The floor of the sales ring should be of rough concrete laid on a firm gravel base. The floor of the sales ring should be covered with sand, sawdust, or wood shavings to keep dust down and provide a solid footing.

Auctioneer's Box

During a sale, the auctioneer's box is a combination office and scale house and should provide space for the auctioneer, recorder, and weighmaster, and for necessary equipment. Some markets also may find it desirable to provide space for radio broadcasts. A stairway should be provided from the box to the catwalks over the yards.

The auctioneer's box should be 9 feet wide and 10 feet deep. The floor of the box should be 3 feet above the floor of the ring and its ceiling about 18 feet above the box floor. This height gives the auctioneer a clear view of the audience. The side of the auctioneer's box facing the sales ring should be a curved counter projecting 3 feet into the ring. The top of the counter should be roughly 18 inches wide (to provide *desk space* for the auctioneer and recorder) and 7 feet above the floor of the sales ring. At markets that handle Brahman cattle the top of the counter should be at least 8 feet above the floor of the sales ring. The box also should afford employees working in it adequate protection from animals in the ring.

The scale beam in the auctioneer's box is on the side adjacent to the scale platform where buyers and sellers can watch weight determinations. Windows of proper size and design give the weighmaster a full view of the scale platform. A public-address system, a ticket-carrier system, and adequate lights complete the essential components of the auctioneer's box.

Seating Area and Balcony

The suggested barn seats 213 people. The seating area, sales ring, and auctioneer's box constitute an auditorium in which the sales ring and auctioneer's box are the stage. The floor of the seating area rises from an elevation of about 12 inches at the ring at the rate of 15 inches for each row of seats.

To provide passage into and out of the barn and standing room space for overflow audiences, an 8-foot balcony directly to the rear of the seats is suggested. About 2 feet of the balcony directly behind the seats and space each side could be used for standing-room for approximately 100 people.

Market Offices and Restaurant

The main entrance in the center of the building leads directly to the lobby which is about 8 feet wide and 12 feet deep. At the rear of the lobby, space is provided for 2 telephone booths, each booth about 3 by 4 feet. The lobby separates the market offices and

ie office space is divided into 2 offices--a general office 17 feet wide and
ide dimensions) of 204 square feet, and a private office for the manager is
t, or 120 square feet. A counter 14 feet long divides the general office into
, a passage about 4 feet wide and 17 feet deep, is for the buyers, sellers,
s; the second about 7 feet wide and 17 feet deep will accommodate the 4 or
employed on sales days. There are 2 entrances to the general offices, one
d the other from the front of the building. The restaurant, 12 feet wide and
i an alcove for a kitchen, can serve 12 people simultaneously. The toilets
ide of the auction barn and directly behind the kitchen. Each toilet is 7 by

JGGESTED LAYOUT FOR A MARKET THAT WEIGHS LIVESTOCK
AFTER THEIR SALE--TYPE 1

;ypes and amounts of facilities needed and their arrangement on a market that
after they leave the sales ring, a market handling about 600 cattle and
is selected. This market tags cattle, mixes owner lots in the same pens,
buyer and seller pens, starts the price of animals, and sells most cattle
are received and shipped by motortrucks.

Facilities Needed

Truck Docks

l market should have 4 truck docks spaces for receiving livestock: Two for
m stake-body type trucks, and 1 for tractor-trailer trucks. This dock should
size, and arrangement shown in figure 28.

ut, 2 truck dock spaces should be provided for pickup trucks. The design of
size and arrangement, should correspond with those shown in figure 28 for
truck docks spaces also should be provided for loading out larger trucks,
ake-body type trucks and 1 dock for tractor-trailer trucks. The design,
ent of these docks should correspond with those illustrated for the 2 larger

Loading Pens

en should be provided for each of the 4 platform spaces on truck loading
would be used to hold lots of animals driven to the truck dock while the
iyer is at the dock. Loading pens also provide space for buyers to *shape up*
proper shipping lots without interference to other operations. Each of the
tains approximately 150 square feet of space or a total of about 600 square
pen has 2 gates--one opening into a chute pen and the other opening into
Gate and fence construction should be of the design shown in figure 32.

Cattle Tagging Chute

One double-lane cattle tagging chute of the design and size shown in figure 31 should be provided.

Catch Pens

Various operations in the receiving and selling cycle are performed in sequence and delays in one operation in the cycle tends to cause equal delays in the operations that follow. In these cycles catch pens break the sequence of the operations and permit their performance at different rates. Catch pens also permit the performance of certain operations with less effort and, in some instances, reduce total labor requirements. Five catch pens are suggested for the market, 4 for receiving and 1 for selling. Each pen should be 10 by 28 feet, a size which will hold about 20 head of cattle for short periods. Total space in catch pens would be 1,400 square feet.

The catch pen for use in the selling cycle should connect directly with a chute pocket, the dimensions of which should be not more than 10 by 10 feet. The chute pocket is used to hold livestock prior to driving them into the feeder chute. Although this chute pocket contains 100 square feet of space, no more than 5 or 6 head of cattle should be held in it at one time because of the difficulty in working large numbers of cattle in a comparatively small area.

The fences and gates of catch pens and the chute pocket should be of the design shown in figure 32. Each catch pen should have 2 gates, each of which will open into different alleys or into an alley and a chute pocket.

Holding Pens

The number, sizes, and types of holding pens needed on livestock auction markets are directly related or are dependent on such factors as: (1) Volume of business handled per sale, (2) species handled, (3) average sizes, by species, of animals handled, (4) penning practices employed, (5) number of buyers and the ranges in volume purchased, and (6) loading-out of animals in relation to hours of sale.

As previously pointed out, the assumed market would handle about 600 cattle and about 150 hogs per sale. Because of seasonal variations at several sales each year the volume consigned may be twice as large as normal. Therefore, allowances must be made for peak volumes. Although the suggested market probably would handle a few sheep, horses, goats, and mules, it is assumed that the volumes handled would be too small to justify specially designed holding pens. Cattle pens can be used for horses, mules, sheep, and goats. Specially designed pens would be needed for bulls, which should be held in individual pens.

Because of practices previously discussed and the relatively large cattle pens found on most auctions, it is rarely possible for markets to utilize more than 80 percent of the total space in buyer and seller holding pens. However, by providing cattle pens of smaller sizes than those now used, or by designing the pens so that one relatively large pen with gates opening on 2 alleys could be converted into 2 pens by use of a center gate, it is estimated that the utilization could be increased to 90 percent of the total space. Utilization of pen space would not be increased for hogs.

In determining the amount of pen space per animal to provide in holding pens, at least 3 factors should be considered: (1) Weight of animals, (2) length of time animals will be held in pens, and (3) the climate of the area.

The average weight of cattle handled by markets in the Southeast that weigh livestock after they leave the sales ring is about 500 pounds. Although the length of time cattle are held in buyer and seller pens is extremely variable, the average total is about 6 hours. Hogs are held in buyer and seller pens about the same length of time. Allowances should be made in the average amount of pen space provided per animal because of high temperatures during the late spring, summer, and early fall when overcrowding will result in heavy losses from deaths and shrinkage.

It is suggested that 14 square feet of pen space per head be provided in both sellers' and buyers' cattle holding pens, and that 6 square feet of space per head be provided in holding pens for hogs.

Table 3 shows the number and types of pens, and the total space needed for holding pens at a market that would handle about 600 cattle and 150 hogs per sale but which on peak-volume sales must accommodate roughly 950 cattle and 230 hogs. Consideration has been given to the possibilities of utilizing a limited number of cattle and hog pens and all the bull pens as both sellers' and buyers' pens. Where holding pens are to be used by both sellers and buyers on the same day, they necessarily must be emptied of sellers' livestock before they can be assigned to buyers. There are limitations as to the number of cattle pens that can be efficiently utilized as dual-purpose pens.

In addition to 8 bull pens, 55 cattle holding pens are suggested for the assumed market: 28 seller pens, 17 buyer pens, 5 dual-purpose pens, and 5 *overflow* pens. Four of the seller pens should be located in the yards so that they can be divided into 8 pens by the use of a middle gate. By using the 5 dual-purpose pens as seller pens before the sale, the market would have a total of 37 seller pens. By using them as buyer pens after they have been emptied, the market would have a total of 22 buyer pens.

A total of 8,890 square feet of pen space is provided in the 28 seller cattle pens and the 5 dual-purpose pens. Allowing 14 square feet per head and with 90 percent utilization of the space, about 572 cattle can be accommodated. The 8 dual-purpose bull pens would accommodate 8 bulls. The catch pen previously suggested for selling operations also would be used as a sellers' holding pen prior to the sale. It would hold about 20 head. Thus, the sellers and dual-purpose cattle pens would accommodate about 600 head.

The 5 overflow cattle pens would be used principally for yarding sellers' cattle which arrive a day or 2 before the sale, buyers' cattle left on the market for a day or 2 after the sale, and as regular holding pens on peak-volume sales days. These pens provide a total of 5,442 square feet of space and would accommodate roughly 350 cattle, assuming 90 percent utilization of pen space. Thus the total capacity of the market for cattle is roughly 950 head. However, the market cannot handle 950 cattle as efficiently as 600 head.

A total of 55 hog pens is suggested for the market. Of this number 45 are seller pens, 4 are buyer pens, and 6 are dual-purpose pens. Before the sale 45 hog pens would be available for sellers. After the sale 10 hog pens would be available for buyers. In addition, buyers' hogs could be yarded after their sale in more of the seller pens. The 40 seller hog pens provide 960 square feet. Allowing 6 square feet per hog, these pens would accommodate about

Table 3.--Types and number of holding pens and total pen area for suggested livestock auction market that practices after-sale weighing of livestock

Type of pen and dimensions	Pens Number	Pen area Square feet
Sellers' cattle pens 10 by 28 feet 1/	2/ 28	7,840
Buyers' cattle pens		
10 by 42 feet	12	5,040
10 by 21 feet	5	1,050
Subtotal	17	6,090
Dual-purpose cattle pens 10 by 21 feet 3/	5	1,050
Overflow cattle pens 4/		
30 by 38 feet	2	2,280
30 by 31 feet	1	930
30 by 52 feet	1	1,560
14 by 48 feet	1	672
Subtotal	5	5,442
Dual-purpose bull pens 3 feet 6 inches by 8 feet 3/	8	224
Total cattle-pen space		20,646
Seller hog pens 4 by 8 feet 5/	30	960
Buyer hog pens		
8 by 12 feet	1	96
8 by 16 feet	3	384
Subtotal	4	480
Dual-purpose pens 3/		
8 by 8 feet	5	320
8 by 12 feet	1	96
Subtotal	6	416
Total hog pen space		1,856
Total holding pen space		22,502

1/ Four of these pens would have a center gate which can be closed to divide the pen into 2 pens.

2/ Includes the 4 catch pens suggested in connection with receiving cattle into the market. When needed, these pens should be used for penning the last seller's cattle arriving on the market.

3/ Dual-purpose pens are pens that might be used alternately by either sellers or buyers.

4/ Overflow cattle pens are pens that would be used primarily as sellers' pens when the volume of cattle consigned per sale exceeds the capacity of the designated sellers' and dual-purpose pens--or roughly 600 head. Overflow pens also would be used for holding cattle received the day before the regularly scheduled sales day.

5/ Fifteen of these pens would have a center gate which can be closed to divide the pen into 2 pens.

160 hogs. The 6 dual-purpose hog pens containing 416 square feet would hold approximately 70 hogs. Thus the capacity of the hog section would be about 230 hogs.

Although it is not suggested that the floors of the cattle pens be paved, adequate provisions should be made for drainage. Shallow open ditches along fence lines are suggested if the topography is suitable. Where natural drainage is not possible an underground drainage system should be provided. The floors of the hog pens should be one continuous slab of rough concrete and should have adequate drainage for flushing and cleaning. An open drain having a top width of 16 inches and a semicircular bottom 8 inches in diameter is suggested. Wooden boards should cover the drain where the alley crosses it. The fences and gates of holding pens for cattle and hogs should be of the designs shown respectively for these species in figures 32 and 33. Bull pen fences should be stronger.

Alleys

Although alleys should be of sufficient width to permit the free flow of all size lots of livestock yarded in holding pens, their width should correspond with the width of adjacent holding pens so that, when completely open, pen gates will span the width of the alley. Furthermore, there should be a sufficient number of properly located alleys to permit a free flow of livestock to or from all sections of the market yards when receiving, selling, and loading out are underway simultaneously within the flow lines of one cycle crossing the flow lines of another cycle.

In the cattle section alleys should be 10 feet wide, except those near bull pens and overflow pens. Where fire lanes are required, the width of alleys must, of course, conform with local ordinances. Ten-foot alleys should permit the free flow of all size lots of cattle yarded in the holding pens, the use of a truck for cleaning and disinfecting, hauling feed to the pens, and other comparable uses.

Generally, each row or tier of holding pens should be separated by a drive alley which leads toward or away from the sales ring. In addition, drive alleys should parallel both outer sides of the yards. Cross alleys, which permit cattle to be driven from one side of the yards to the other, should be placed at intervals to provide a free flow of cattle through the yards without out-of-line and back drives when 3 cycles of operations are being carried on simultaneously.

Alleys in the bull-pen section should be about 4 feet wide. All alleys in the cattle section should be paved either with concrete or black-top on a stabilized base. In the hog section alleys should be 4 feet wide. Drive alleys should again separate each row or tier of pens and cross alleys should be placed at needed intervals. All alleys in this section should be paved with concrete.

Block Gates

As previously pointed out, gates of holding pens should be wide enough to block the alley when opened. Block gates also should be provided at the intersections of all alleys and at strategic points between these intersections. They are used for diverting the flow of livestock and for creating pens for temporarily holding animals in the alleys. Block gates should be the same height as the pen gates and fences in the respective sections of the market.

Block gate construction should be similar to that of holding-pen gates (figs. 32 and 33).

Feeder Chute

The feeder chute suggested for driving cattle singly or in lots into the sales ring is of the design and size shown in figure 36. The chute should be covered by the yard roof and its floor should be of rough concrete.

Scales

A weighbeam-type scale with a 10,000-pound capacity, a 5-pound minimum graduation, a printing attachment, and a 7- by 14-foot platform with racks is suggested (fig. 34).

The scalebeam should be arranged so that the weighmaster will face the platform when weighing animals and be in position to observe both the drive-on and the drive-off ends of the platform so that he can see when the gates are closed properly. The weighmaster also should be in a position to receive sales tickets from the recorder readily. Sellers and buyers also should be able to observe weight determinations and recordings.

Catwalks

Catwalks should be constructed over the yards to provide market patrons, supervisory employees and others a means of access to various parts of the yards for observing livestock without the necessity of walking through alleys. Approximately 660 feet of catwalks would be needed to provide a complete loop over the yards and provide connections from this loop to the truck docks and to the auction barn. The suggested design of catwalks is shown in figure 40.

Water Troughs, Hayracks, and Grain Troughs

Although the number of cattle holding pens in which water troughs, hayracks, and grain troughs are provided should be determined by practices in specific localities, as a minimum these facilities should be provided in the 5 overflow pens. Two 10-foot water troughs and one 20-foot combination hayrack and grain trough should be placed in each of these pens, or a total of 100 linear feet of water troughs and 100 feet of combination troughs. The designs for the facilities are shown respectively in figures 37 and 39.

One water trough running the full depth of the pens should be provided in each of the 4 buyer and every other one of the dual-purpose hog pens, for a total of 84 feet of hog water troughs (fig. 38). In addition, adequate water hydrants with threaded hose connections also should be provided in the hog section for wetting down and washing out pens.

Roof

With the exception of the overflow cattle pens, loading pens, and truck docks, the entire market yard should be covered. The monitor-type roof is suggested, with eaves about 12 feet above the ground level. The roof should be constructed of either corrugated metal or prepared asphalt-type roofing. About 31,965 square feet of roofing would be needed.

Yard Lights

As night work frequently is necessary, it is suggested that the yards be adequately lighted. The light system should consist of a string of 100-watt lights with shallow reflectors over the center of each row of pens and over each alley. Lights over the pens should be spaced at 40-foot intervals and those over the alley at 20-foot intervals. The lights should be at least 12 feet above ground and well insulated against the weather. Lights also should be provided over all truck docks.

Kitchen Pen

A *kitchen pen* is a small holding pen located directly to the rear of the auctioneer's box and between the 2 main drive alleys connecting the yards with the sales ring. It is used for temporarily holding animals that enter the ring out of turn or animals which cannot be identified by ownership. Its use eliminates the necessity for driving animals back into the yards. The kitchen pen should be 9 feet wide and 15 feet deep and contain 135 square feet of space. The pen would hold about 9 cattle. Fence and gate construction should be comparable to that used for cattle pens.

Auction Barn

The auction barn, which should include the sales ring, auctioneer's box, seating area, market offices, restaurant, and toilets, should be of the design and size shown in figure 41.

Ticket Carrier System

A ticket carrier system, connecting the auctioneer's box in the auction barn and the market office, is necessary for conveying the sales tickets to the office after the sale of each lot. The dual overhead wire ticket carrier system is suggested because of the comparativel low initial investment required and the relatively small operating cost.

Market Driveways

The principal market driveway, connecting the auction barn with the public highway and providing access to the aprons adjoining the docks and unloading pens, should not be less than 40 feet wide and should be paved so as to stand up under heavy traffic. Aprons or driveways connecting with the docks and unloading pens should not be less than 100 feet wide and should have a rolled gravel surface.

Parking Areas

Well-defined *off-street* parking areas should be provided on the market for a total of about 210 motor vehicles. These areas should be out of traffic lanes but easily accessible from them. The parking areas should be gravel surfaced, well drained, and with plainly marked individual spaces 10 feet wide.

Arrangement of Facilities

A layout for the entire market is shown in figure 42. The auction barn faces and is connected with a public highway by a 40-foot paved driveway. This driveway enters the market site on one side, curves around to the front of the barn and leaves on the opposite side. The curved driveway connects with 100-foot service driveways on each side of the auction barn which in turn provide access to each side of the yards. Parking spaces for 210 vehicles are provided within the semicircle of the driveway in front of the barn and along each side of the market site opposite the barn and yards.

The sales ring of the barn connects with the 2 alleys used respectively for moving animals up to the ring for sale and to pens after the sale. The multiple-lane feeder chute provides this connection from the incoming alley, the scale platform pen (rack) to the outgoing alley. The seating area and auctioneer's box also connect with the yards by means of steps from the catwalks.

In the layout shown, catch pens and sellers' cattle holding pens have been grouped in 3 rows along the left side of the market yards (the left side when the viewer faces the yards from the auction barn) where livestock would be received. Buyers' cattle holding pens and dual-purpose pens have been grouped in 2 rows on the opposite side of the yards where livestock are loaded out. Part of the row of pens directly to the rear of the sales barn and near the center of the yards are dual-purpose pens which may be assigned either to buyers or sellers. Overflow cattle pens are in a row across the rear of the market yards. The bull pens also are located in this area. With this pen arrangement the flow of cattle generally would be from the left to the right side of the market.

Hog holding pens are grouped in 2 rows at the front of the market near the sales ring. The outside row is seller pens and the inside row is buyer and dual-purpose pens.

The 5 rows of *regular* cattle pens are separated by and open into alleys running from the overflow pens at the rear of the yards to the hog section or the scale platform at the front of the yards. The 4 interior alleys separating the rows of cattle pens are paralleled by an alley on each of the outer edges of yard, so that each pen opens into 2 different alleys-- one on each side. Cross alleys are provided at the rear and midway of the yards. Additional cross alleys can be provided as required by opening all pen gates and blocking the drive alleys in any consecutive series of pens across the yards.

Drive alleys are numbered from 1 to 6, inclusive, beginning at the left side of the yards. Cross alleys are numbered 7 and 8 beginning at the center of the yards. Alley number 3 is the main drive alley used to bring up cattle to the ring. Alley number 5 is the pen back alley.

The 2 rows of hog pens also are separated and flanked on each side by drive alleys leading toward the sales ring. With the exception of the buyer pens which open into only 1 alley, all hog pens open into 2 alleys. In addition to the 3 drive alleys, 2 cross alleys are provided--1 adjacent to and leading into the feeder chute pocket at the sales ring and 1 midway of hog pen section.

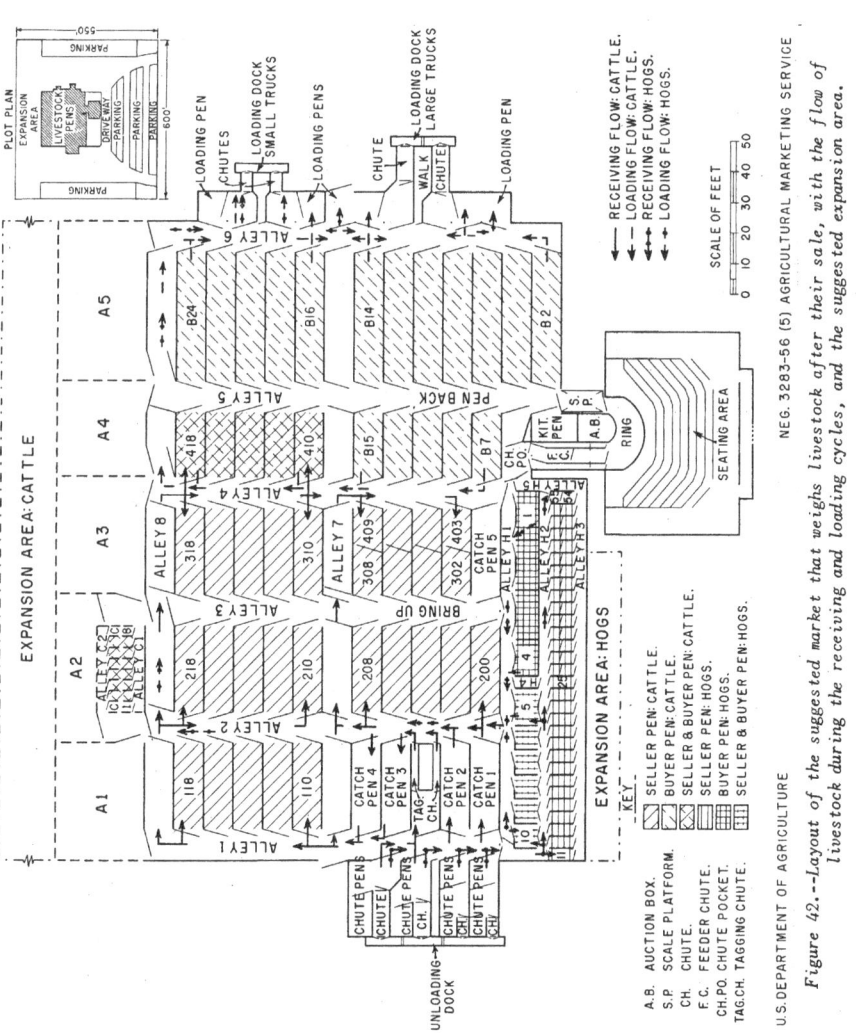

Figure 42.—Layout of the suggested market that weighs livestock after their sale, with the flow of livestock during the receiving and loading cycles, and the suggested expansion area.

Truck docks for unloading livestock are on the left side of the market. This location should hold labor requirements for receiving livestock to a minimum. The docks should project outward from, rather than being enclosed by, the yards so that holding pens and alleys could be expanded without the necessity of moving the docks.

With but one exception, alleys in the hog pen section are 4 feet wide. The interior alley separating the cattle and hog sections is 5 feet wide. Buyer and dual-purpose pens open into the latter alley. Five-foot gates, which are less than the full pen width, open into this alley. Seller pens, which have full-width 4-foot gates, open on each side into 4 alleys.

Drive alleys are numbered H 1 to H 3, inclusive, beginning with the interior 5-foot alley. Cross alleys are H 4 and H 5. Alleys number H 2, H 3, and H 5 are the bring up alleys. Penning back is through alley number H 1. The 2 truck docks for loading-out livestock should be on the right side of the yards. The location of truck docks on one side of the yards for receiving and on the opposite side for loading out should permit a direct flow of livestock into and through the market and minimize back drives and out-of-line drives. Docks for loading pickup trucks should be located toward the rear of the market because animals purchased by small buyers usually are yarded after their sale in the holding pens at the rear of the market. The drive of small lots from these pens to the docks would be comparatively short. Truck docks for loading tractor trailers and farm stake-body trucks are located about midway of the depth of the yards. This location should provide the shortest drive from holding pens to truck docks.

Catch pens are located directly across a drive alley from and connect with the chute pens of the truck docks for receiving livestock. Two catch pens are on each side of the tagging chute. These pens should break the sequence of tagging and yarding operations. These pens also should permit workers yarding livestock in sellers' pens to drive large groups of cattle to the seller holding pens instead of comparatively small individual lots as they arrive on the market. These 4 catch pens also would be available for use as holding pens for the last few loads of cattle arriving on the market. One catch pen is located near the sales ring to enable workers bringing-up animals for sale to maintain an adequate number near the ring.

The tagging chute is directly across the alley from the chute pens of the docks for receiving livestock between the 4 catch pens used for receiving. This location provides the shortest drive of cattle from the chute pens to the tagging chute. The tagging chute is as centrally located as possible to all the cattle holding pens in the yards to minimize the length of the drive from this chute to the pens.

The scale platform should be located between the exit gate of the sales ring and the main drive alley leading to the buyer holding pens so that one scale gate forms a part of the sales ring and the other opens on the alley. This location places the scale alongside the auction box and permits the scalebeam to be placed inside the box. This location also permits the workers driving livestock from the sales ring to open and close the on gate of the scale platform and the workers driving livestock from the scale platform to drive them down an alley without a right-angle turn.

Loading pens are adjacent to each of the 4 spaces on the 2 docks for loading-out livestock. Each pen opens on the same drive alley.

To expand or increase the size of the yards, additional cattle pens and alleys would be constructed at the rear. As a first step in expanding cattle holding pens, the 5 large pens at the rear of the yards would be divided into smaller pens, and new overflow pens added. This would increase the depth of the yards, with no increase in width. By expanding in that direction, the same flow lines would be maintained and truck docks would remain in their present location. Bull pens would be expanded by dividing adjacent space in the overflow pens into the smaller pens. Hog holding pens would be expanded by adding the row of pens parallel and adjacent to alley H 3 at the front of the yards. Alley H 6 would be added. The new pens would be 4 by 8 feet with 4-foot gates opening onto alleys H 3 and H 6.

Estimated Cost of Construction

Cost estimates shown on page 70 are based on general construction cost in the Southeast during 1954. These estimates are presented only as a guide for use by market operators in estimating the total market cost and the prospective investment that might be required to construct a facility of the kind and size suggested. These estimates are not intended to replace the estimates of local contractors at the time and place of construction. These estimates are for a market of a size to handle 600 cattle and 150 hogs per sale.

Amount of Land Needed for Market Site

The market layout shown in figure 42 is on a site 600 feet wide and 550 feet deep, and contains roughly 7.5 acres. This width and depth cannot be materially decreased if the suggested market layout is to be maintained and provisions are made at the outset for market expansion. Obtaining the necessary depth required for this layout may necessitate the buying of more *frontage* than is needed. However, the reverse usually will be true, particularly where sites are along important highways. The cost of the land, placed in condition to build, must be added to the estimated cost of constructing the facilities to determine the total market cost. Because of variations in land costs no estimates are made in this report.

How Suggested Facility Should Operate

The various components of the market are arranged so that definite flow lines must be observed during the 3 major cycles of operations, if the facility is to operate efficiently. These lines are shown by flow diagrams (figs. 42 and 43). Other flow lines would be possible by swinging some of the pen gates differently. Market operators who revise the basic layout to fit a special tract of land, or for other reasons, should determine which lines of flow will be most efficient before deciding on the arrangement of the various components of the market.

Receiving Operations

Cattle unloaded at the truck docks from farm stake-body and tractor-trailer trucks are driven from chute pens across alley 1 into the tagging chute where they are tagged and the receiving ticket is prepared. The cattle are then driven from the tagging chute into either catch pen 3 or 4. Cattle arriving on pickup trucks are tagged and the receiving ticket prepared while the animals are on the truck. After they are unloaded these cattle are driven

Item	Estimated Cost Dollars
Docks, chutes, and chute pens.	1,760.00
Cattle pens:	
Fences and gates 5 ft. high, five 2- by 6-inch rails, 6- by 6-inch and 8- by 8-inch posts	
4,265 lin. ft. @ $2.40 per ft.	10,236.00
Hayracks and feed troughs	
100 lin. ft.@ $2.00 per ft.	200.00
Water troughs	
10 troughs @ $27.50 each.	275.00
Paving alleys (black top)	
1,232 sq. yds. @ $1.90 per sq. yd.	2,340.80
Hog pens:	
Fences and gates, 3 ft. 6 in. high, two 1- by 10-inch rails six 1- by 6-inch rails, 4- by 4-inch, and 6- by 6-inch posts	
1,056 lin. ft. @ $2.65 per ft.	2,798.40
Water troughs	
84 lin. ft. @ $2.20 per ft.	184.80
Paving pens and alleys (concrete)	
400 sq. yds. @ $2.20 per sq. yd.	880.00
Roof over yards	
31,965 sq. ft. @ $0.50 per sq. ft.	15,982.50
Catwalk over yards, including 2 stairways	
660 lin. ft. @ $2.75 per ft.	1,815.00
Yard lighting	550.00
Feeder chute, multiple type.	330.00
Auction barn, wood frame construction	
Includes stairs on front, auction box, offices, restaurant and toilets	
3,060 sq. ft. @$3.30 per sq. ft. 1/	10,098.00
Scale, 7- by 14-ft. platform	2,420.00
Seats, 213, theatre-type @ $5.50 each	1,171.50
Public address systems, 2 systems@ $550 each	1,100.00
Ticket carrier system	165.00
Water lines in cattle and hog pens, cafe, and toilets	1,320.00
Sanitary sewers (inside market area)	1,980.00
Paving (rolled gravel):	
Driveways, aprons, and parking	
18,100 sq. yds.@ $0.55 per sq. yd.	9,955.00
Subtotal	65,562.00
Engineering fees @ 6 percent	3,933.72
Total cost of market	69,495.72

1/ Estimate includes all electrical work, lighting fixtures, plumbing and toilet facilities, but does not include office and restaurant equipment.

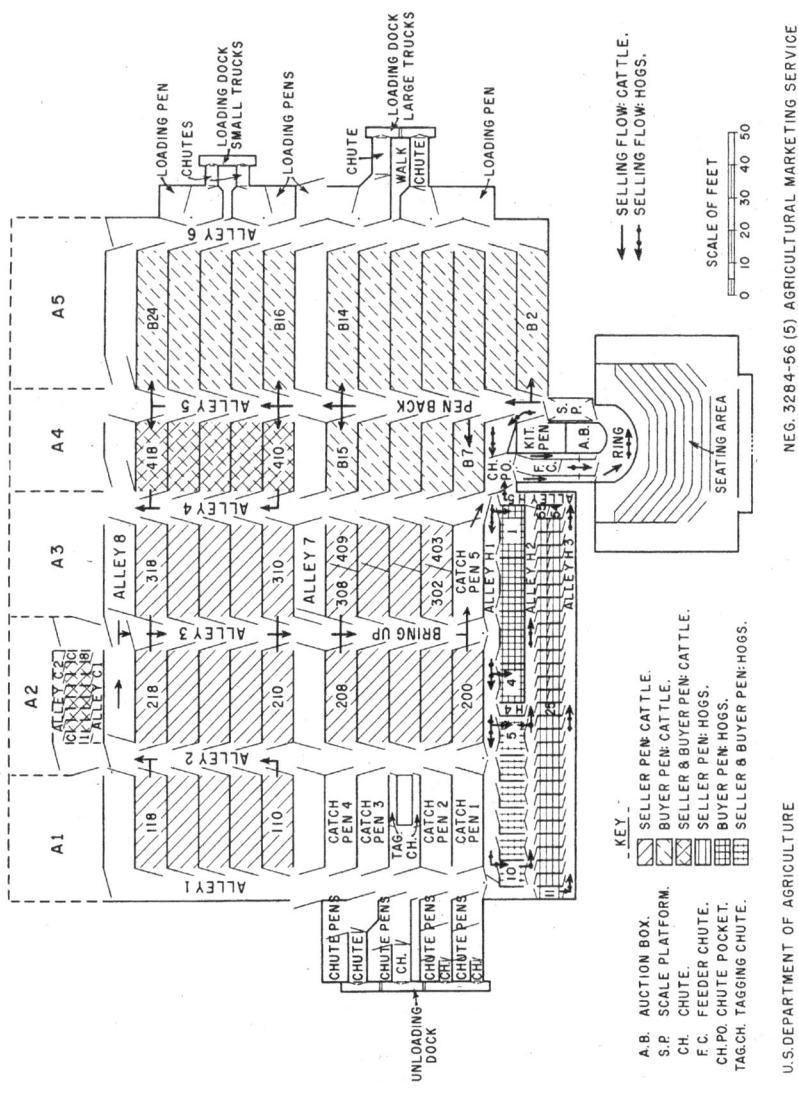

Figure 43.--*The flow of cattle and hogs during the selling cycle on the suggested market that weighs livestock after their sale.*

into either catch pen 1 or 2. When one of the catch pens has been filled, all animals in the pen are driven to and penned in a seller holding pen.

Small lots of cattle, regardless of their time of arrival, are yarded in the small seller pens, 302 to 308 or 403 to 409. Cattle in catch pens 3 and 4 are driven to these pens through alleys 1, 8, and 4. Those in catch pens 1 and 2 are driven to these pens through alleys 2, 7, and 4. The dual-purpose pens, 410 to 418, are filled with sellers' cattle as early as possible and they are the first cattle sold so that the pens will be available to buyers. Cattle in catch pens 3 and 4 are driven through alleys 1, 8, and 4 and those in catch pens 1 and 2 would be driven through alleys 2, 8, and 4 to reach the dual-purpose pens. Although there are several possible routes from the catch pens to other seller pens, generally cattle from catch pens 1 and 2 are yarded in pens 210 to 218 and 310 to 318 by way of alleys 7 and 3. Cattle from catch pens 3 and 4 are yarded in the same pens by way of alleys 1, 8, and 2 or alleys 1, 8, and 4. Cattle from all 4 catch pens would be driven directly across alley 2 to pens 200 to 208. Cattle from catch pens 3 and 4 would be driven through alley 1 to pens 110 to 118.

As hogs are yarded by ownership, each lot unloaded is driven directly to the sellers' holding pens. The first sellers' holding pen to be filled depends on the number and size of each lot. Pens H 26 to H 55 are designed for small lots common to most markets. Pens H 11 to H 25 are designed for medium size lots, and dual-purpose pens H 5 to H 10 are designed for larger size lots. Hogs yarded in dual-purpose pens H 5 to H 10 are among the first lots to be sold so that these pens will be available for buyers. Hogs yarded in pens H 11 to H 55 and H 5 to H 10 are driven up alleys 1 and H 2 to the assigned pen.

Selling Operations

The movement of animals on the market during the sale is shown in figure 43. As the flow lines indicate, operations in connection with the sale involve the most extensive movement of animals required at any stage of the marketing process. All operations in this cycle must be keyed to the rate of sale.

Hogs are the first species sold, and animals yarded in the dual-purpose pens H 5 to H 10 lead off. Hogs are driven out of these pens into alley H 2 then into alley H 5, through this alley and into the center lane of the feeder chute. Hogs yarded in pens H 11 to H 55 are driven up alleys H 3 and H 5 into the feeder chute. In the chute animals are sorted into salable lots and driven into the sales ring. From the sales ring they are driven onto the scale platform and after they are weighed are driven through alleys 5 and H 1 to the assigned buyer pens.

Cattle in the dual-purpose pens, 410 to 418, are the first to be sold. From these pens cattle move through alley 4, 8, and 3 to catch pen 5. From this pen cattle are driven by small lots into the chute pocket. If they are sold singly cattle are driven from the chute pocket into the side lanes of the feeder chute and then into the sales ring. If sold in groups, they are driven through the center lane of the feeder chute into the sales ring. After they are sold cattle are driven from the sales ring onto the scale platform for weighing, and then down alley 5 to the assigned buyer pens. As soon as pens 410 to 418 are empty they are used for yarding buyers' cattle. Cattle yarded in catch pens 1 and 2 are driven into alley 2, down alley 2 to alley 7 and alley 3 and up alley 3 to catch pen 5. Cattle yarded in catch pens 3 and 4 are driven into alley 1 through alley 8 to alley 3 and up

3 to catch pen 5. Cattle yarded in pens 200 to 218, 302 to 318, and 403 to 409 are
n through alley 3 to catch pen 5. Alley 3 is the main drive alley used to bring up
e to the sales ring. Alley 5 is the main pen-back alley.

A major consideration in obtaining efficiency in sales operations is the proper
gement of workers in the sales ring and the auctioneer box. An arrangement of workers
has proved desirable on a number of markets is to have 1 ringman stationed at the *in*
of the sales ring and 1 at the *out* gate; the starter in the ring near the *in* gate; the
der at the counter in the auctioneer's box near the *in* gate to the ring; and the auc_
er next to or at a position almost directly in the center of the box where he can command
ttention of the bidders and have an unobstructed view of the sales ring. The weighmaster
ationed in the box on the side near the *out* gate of the sales ring, just to the rear of
uctioneer so that he can receive the scale tickets from the recorder.

The ringman allows animals to enter the sales ring by opening the *in* gate. The starter,
ed near this gate, is in a position to make a rapid appraisal of its value and announce
rice to the auctioneer. The auctioneer begins the sale at the starter's price and con-
s his chant while the animal is displayed.

The recorder picks up the number of the animal from its tag as it enters the sales ring,
e gateman may call the number of the animal to the recorder. Animals are driven out of
ing after the auctioneer announces the price paid and the name of the buyer. The
der records the price and the buyer's name on the scale ticket and passes it to the
master.

Loading-Out Operations

The flow diagram of loading-out operations is shown in figure 42. The physical handling
ved in loading-out operations requires fairly short and direct drives of animals in
ively large lots. Cattle yarded in pens B 2 to B 24 are driven into and through alley 6
e truck docks. Animals yarded in pens B 7 to B 15 and 410 to 418 are driven into
s 4, 8, and 6 to the docks. Alley 5, the pen-back alley, is not used for loading-out
e until after the sale.

Hogs are driven through alleys H 1, 2, 8, and 6 to the docks if loaded out before the
is over. After the sale they are driven through alleys H 1, 2, 7, and 6 to the docks,
ey may be driven through alley H 1 and 1 and loaded out at the same docks at which they
received. Both cattle and hogs may be loaded out from these docks after the sale because,
the operation of only one cycle being carried on, the problems of mixup and delays from
flows are avoided.

Labor Requirements

Nineteen workers, other than the office crew, would be required to operate the improved
t or 3 fewer workers than required for the typical market. The crew consists of the
oneer, starter, weighmaster, recorder, 2 ringmen, the yard foreman, and 12 yardmen. The
tion in the number of workers required is in the yardmen and is made possible by
ved facility designs and layout.

Nine workers directed by the yard foreman receive livestock before the sale starts, or one less than required at the typical market. Two workers prepare the receiving tickets, 2 workers tag, and 5 workers unload and pen livestock in seller pens. One worker is primarily responsible for yarding hogs in seller holding pens in conjunction with his work in receiving cattle. The reduction in the receiving crew is in the number of workers required to unload and pen livestock. When the hog sale starts 6 workers are transferred to selling operations. When the hog sale is over an additional worker is transferred to the cattle selling operation. The 2 remaining workers perform all receiving and loading-out operations during the sale.

The hog selling crew consists of 15 workers: The auctioneer, recorder, starter, weighmaster, 2 ringmen, and 9 yardmen. A yard foreman directs the yardmen. Of the 15 workers 6 are from the receiving crew and 9 report for duty when the sale starts. The number of workers is 2 fewer than is required by the typical market. The arrangement of the hog pens in relation to the sales ring decreases the number of workers required.

The cattle selling crew consists of 16 workers: The auctioneer, recorder, starter, weighmaster, 2 ringmen, and 10 yardmen. A yard foreman directs the yardmen as well as the workers receiving and loading out during the selling operations. The additional worker in the cattle selling crew is obtained from the receiving crew. The number of workers is 3 fewer than is required by the typical market.

The feeder chute and its relation to the sales ring makes it possible for 3 workers to drive cattle into the sales ring. Four workers perform this job on the typical market. The straight away alley from the sales ring to the cattle buyer pens, and the 10-foot-wide buyer cattle pens, make it possible to yard animals after the sale just as quickly and efficiently and with 1 less worker. Two scalemen are required for the improved market, compared with 3 for the typical market. This reduction is made possible by locating the scale platform adjacent to the sales ring. The number of workers needed to check animals in the buyer pens and assist in loading-out livestock immediately after the sale is 10, or 2 fewer than is used at the typical market because pens are accessible from either side and adequate block gates and space in alleys are available.

SUGGESTED LAYOUT FOR A MARKET THAT SORTS AND WEIGHS HOGS ON ARRIVAL AND WEIGHS CATTLE BEFORE THEIR SALE--TYPE 2

A market having a capacity for about 1,260 hogs and 260 cattle per sale is selected to show an improved layout where hogs are weighed and sorted by *grades* on arrival and cattle are weighed before they enter the sales ring. Other practices on this market which affect its size and layout are: (1) Owner's lots of cattle are mixed in the same seller pens, (2) cattle are tagged, (3) both buyer and seller pens are provided, (4) graded hogs are sold while they are in holding pens, and (5) a high proportion of cattle are sold singly.

Facilities Needed

Truck Docks

One truck dock of the size and design shown in figure 28 should be provided for both receiving and loading-out hogs. The 3-level dock should have 4 platform spaces, chutes, and chute pens--2 for pickup trucks, and 1 each for farm stake-body trucks, and for tractor-trailer trucks. In addition a double-deck truck dock (fig. 30) should be provided exclusively for loading-out hogs. One truck dock containing platform space, chutes, and chute pens for handling 3 trucks simultaneously should be provided for receiving and loading-out cattle. One dock should be provided for each type of truck. These docks should be constructed in accordance with the design shown in figure 28 except that 1 of the 2 spaces for pickup trucks should be eliminated. This dock would be 33 feet wide rather than the 47 feet recommended when 4 spaces are provided.

Loading Pens

Two loading pens each 7 by 30 feet are suggested at the double-deck truck dock for loading-out hogs. These pens would contain a total of 420 square feet and would have a total capacity of about 80 hogs. As some hogs would be loaded out at the same truck docks at which they were received, the chute pens at the docks would serve as loading pens. Cattle also would be loaded out at the same docks at which they were received. Therefore catch pens used in receiving and the dock chute pens would serve as cattle loading pens. Fence and gate construction are shown in figures 32 and 33.

Facilities for Sorting Hogs into Market Grades

Facilities for sorting hogs should consist of a sorting alley, 4 sorting pens, a scale platform, and scale house. The overall dimensions of the area required for the facilities are about 22 feet by 30 feet. The sorting alley, 22 feet long and 8 feet wide, parallels a chute alley. Three 5-foot gates open from the chute alley into the sorting alley. Four sorting pens, each 4 feet wide and 7 feet deep open on one side into a sorting lane constructed in one end of the sorting alley. The sorting lane is 2 feet wide and 10 feet long. Hogs must move single file through this 2-foot lane and are driven from it into the proper *grade* pen by partially opening the 4-foot pen gate into the lane. Five-foot block type gates also should be provided at each end of the sorting lane. Each sorting pen is designated for a specific market grade and each pen has a holding capacity of about five 200-pound hogs. On the side opposite the sorting alley and lane, the sorting pens open through full-width 4-foot gates into a 4-foot alley adjacent and parallel to the scale platform. Two 4-foot gates lead from this alley onto the scale platform so that hogs can be driven onto the platform from either direction in the alley. The scale platform is 14 feet wide and 7 feet deep. Two 5-foot gates open from the scale platform onto a hog drive alley. All fences and gates in this area except those bordering the chute alley should be of the design shown in figure 33. The scale house, which completes the facilities in this area, is 18 feet wide and 6 feet deep. It should be of frame construction with windows of adequate size, properly located to permit the weighmaster to have an unobstructed view of the scale platform.

Catch Pens and Scale Pocket

A series of 3 catch pens should be provided in the hog section of the market, to sort and to hold small salable lots of hogs near the sales ring. The first catch pen should contain about 64 square feet of space. It would be used to sort odd lots of hogs into salable lots. Each of the next 2 catch pens should contain about 24 square feet of space.

Five catch pens and a scale pocket should be provided on the cattle section of the market. Three catch pens each 10 by 21 feet and containing a total of 630 square feet of space would be used for receiving cattle. One catch pen containing 210 square feet would be used to hold cattle adjacent to the scale pocket near the sales ring. Its capacity for short holds is about 15 head of cattle. A scale pocket, 10 by 10 feet, should be provided next to the scale platform to hold small lots of animals for comparatively short periods. Cattle would be driven from this pocket onto the scale platform singly or in small lots. The fifth catch pen, which should be 9 feet wide and 10 feet deep would be used to hold single or small lots of animals that have been driven out of the sales ring after their sale but prior to their assignment to a buyer's pen.

Holding Pens

The suggested market would handle about 1,260 hogs and 260 cattle per sale. The amount of space provided in the hog holding pens should be sufficient to hold the largest number of hogs received for any sale during the year. Moreover, special water facilities must be provided in all hog pens. Thus, part of the pen space for hogs would be unoccupied during many sales throughout the year. In determining the total amount of space needed in cattle holding pens, provisions must be made for handling peak volumes. Relatively large *overflow* pens for cattle usually will suffice.

Although the suggested market might handle a few sheep, goats, horses, and mules, the volumes handled would not justify the construction of specially designed pens. Cattle pens would be used for holding horses, mules, sheep and goats. Specially designed individual pens would be needed for bulls.

As previously mentioned, it is rarely possible to utilize more than 80 percent of the space in buyer and seller pens at present. However, by providing cattle pens smaller than those now used and by designing pens so that one relatively large pen with gates opening on 2 alleys can be converted into 2 pens by use of a center gate, space utilization could be increased to about 90 percent. As the same penning practices currently used in yarding hogs would be used on the suggested market, the utilization of pen space would not be increased.

The average weight of cattle handled at auctions in the Southeast is about 500 pounds. Cattle would be held in pens an average of about 6 hours. Based on these factors and the relatively high temperatures in the Southeast during much of the year, it is recommended that 14 square feet of pen space per head be provided in both buyer and seller cattle pens. Based on an average weight of 200 pounds and a holding period of about 6 hours, 5 square feet of pen space per head is suggested for pens for *graded* hogs. However, as *odd-lot* hogs are yarded by ownership, 6 square feet of pen space per animal is recommended for these lots.

Table 4 shows the types of pens, the number of pens of each type, and the amount of space needed for a *graded-hog* market that would handle 1,260 hogs and 260 cattle per sale. On peak volume sales days the market could accommodate about 520 cattle. Consideration has been given to the possibilities of utilizing a number of cattle pens and all the bull pens as dual-purpose pens. As dual-purpose pens must be emptied of sellers' cattle before the pens can be assigned to buyers, there are limitations as to the number of pens that can be efficiently utilized by a market handling only 260 cattle per sale. A total of 22 cattle holding pens is suggested; 12 seller pens, 5 buyer pens, and 5 dual-purpose pens. Nine of the sellers'

Table 4.--Types and number of holding pens and total pen area for suggested market that practices sorting hogs and presale weighing of cattle

Type of pen and dimensions	Pens Number	Pen area Square feet
Sellers' cattle pens		
10 by 28 feet 1/	6	1,680
10 by 21 feet 2/	6	1,260
Subtotal	3/ 12	2,940
Buyers' cattle pens		
10 by 28 feet	5	1,400
Dual-purpose cattle pens 4/ 10 by 21 feet	5	1,050
Overflow cattle pens 5/		
30 by 48 feet	1	1,440
30 by 52 feet	1	1,560
30 by 38 feet	1	1,140
Subtotal	3	4,140
Dual-purpose bull pens 4/		
8 feet by 3 feet 6 inches	6	168
Total cattle pen space		9,698
Graded hog pens		
10 by 42 feet	10	4,200
10 by 21 feet	2	420
9 by 20 feet	2	360
Subtotal	14	4,980
Sellers' hog pens		
4 by 8 feet 6/	27	864
4 by 6 feet	4	96
Subtotal	31	960
Buyers' hog pens		
10 by 19 feet	2	380
10 by 15 feet	2	300
Subtotal	4	680
Dual-purpose hog pens 4/ 8 by 15 feet	5	600
Total hog pen space		7,220
Total holding-pen space		16,918

1/ These pens would have a center gate which can be closed to divide the pen into 2 pens.

2/ Three of these pens would have a center gate which can be closed to divide the pen into 2 pens.

3/ Includes 3 catch pens suggested in connection with receiving cattle into the market. When needed these pens should be used for penning the last of the seller's cattle arriving on the market.

4/ Dual-purpose pens are pens that might be used alternately by either sellers or buyers.

5/ Overflow cattle pens are pens that would be used primarily as sellers' pens when the volume of cattle consigned per sale exceeds the capacity of the designated sellers' and dual-purpose pens, or roughly 260 head. Overflow pens also would be used for holding cattle received the day before the regularly scheduled sales day.

6/ Thirteen of these pens would have a center gate which can be closed to divide the pens into 2 pens.

pens should have middle gates. By using the 5 dual-purpose pens as seller pens before the sale, the market would have a total of 26 for sellers' cattle. By using the dual-purpose pens as buyer pens after they are emptied of sellers' cattle, the market would have 10 buyer pens.

A total of 3,990 square feet of pen space is provided in the 12 seller cattle pens and the 5 dual-purpose pens. Allowing 14 square feet per head, and with 90 percent utilization of pen space, about 256 cattle can be accommodated. Six dual-purpose bull pens brings the total to about 262 head.

The 3 overflow cattle pens would be used principally for yarding sellers' cattle which arrive a day or two before the sale, for holding buyers' cattle left on the market for a day or two after the sale, and for peak volume sales. These pens provide a total of 4,140 square feet of space and would accommodate roughly 265 cattle, assuming 90 percent utilization of pen space.

A total of 40 pens should be provided for *odd-lot* or ungraded hogs. Thirty-one of these pens are seller pens, 5 are dual-purpose pens, and 4 are buyer pens. As 13 of the seller pens have middle gates by which they can be divided into 26 pens, these pens plus the 5 dual-purpose pens would provide a total of 49 pens for sellers' hogs. By using the 5 dual-purpose pens after the sale as buyer pens the market would have a total of 9 pens for buyers.

The floors and drainage for cattle and hog pens are the same as those previously described for type 1 market.

An overhead water spray system should be provided for all hog pens for cooling animals during hot weather. This system should consist of adjustable lawn-type sprinkler heads, spaced 15 to 20 feet apart. They should be 10 feet above the pen floor level and connected to a water line having not less than 10 pounds' pressure. The pipe sizes used should vary from $\frac{1}{2}$ to 1 inch, depending upon the number of spray heads and the length of the line. Valves for regulating and controlling the flow through the sprinkler head should be provided.

Alleys

There should be sufficient alleys of adequate width and properly located to expedite the flow of livestock to and from all sections of the market during the receiving, selling, and loading-out without flow lines of one group of operations crossing those in other groups of operations.

Specifications for alleys in the cattle sections are the same as those previously given for a type 1 market. In the hog section alleys should range in width from 4 to 10 feet depending on their use. These alleys also should be paved.

Facilities for Driving Livestock into Sales Ring

The feeder chute, recommended for markets that weigh livestock after they are sold, is not recommended for *graded-hog* markets. Most markets of the latter type are in areas where cattle are weighed before they enter the sales ring and the scale rack takes the place of the feeder chute.

Scales

In addition to the scale suggested in connection with the facilities for sorting hogs by market grades, a scale will be needed for weighing cattle before they enter the sales ring. Both scales should be the weighbeam type having 10,000-pound capacity, a 5-pound minimum graduation, printing attachment, and 7- by 14-foot platform. The rack for the scale used in sorting hogs should be the type shown in figure 35. The platform rack on the scale used for weighing animals before they enter the sales ring should be of the design shown in figure 34.

Roof

The entire market yards should be covered. The monitor type roof described for a type 1 market is suggested. About 30,605 square feet of roofing would be needed.

Other Yard Facilities

A double-lane tagging chute of the size and design shown in figure 31 should be provided. This chute is 10 feet wide and 15 feet deep.

Block gates and lights similar to those described for type 1 markets should be provided. About 600 feet of catwalks would be needed to provide a loop over the market yards (fig. 40). The *kitchen pen* should be 10 feet wide and 12 feet deep. This size pen would have a capacity for about 8 cattle.

In the cattle section, watering and feeding facilities should be provided in at least 3 of the overflow pens. Two 10-foot watering troughs and one 25-foot combination hayrack and grain trough should be provided in each of these pens, for a total of 60 feet of water troughs and 75 feet of combination hayrack and grain trough (see figures 37 and 39 for suggested designs).

In the hog section of the market one water trough, running the full depth of the pen, should be provided along the alternating fence lines of the graded hog pens. Two pens would be served by one trough. A total of 250 linear feet of watering trough would be needed in the graded hog pens. In the section of the market for odd-lot hogs, water troughs are suggested in each of the buyer hog pens and dual-purpose pens. A total of 80 linear feet of troughs would be needed. In addition, adequate hydrants with threaded hose connections should be provided in the hog section for wetting down and washing out the pens. A design for hog troughs is shown in figure 38.

Auction Barn

In addition to the sales ring, auctioneer's box, seating area, market offices, restaurant, and toilets, the auction barn should have auxiliary facilities as a public address system and a ticket carrier system (fig. 41). However, this layout should be modified to provide a second entrance into the sales ring. This entrance can be obtained by converting the ground floor passageway on the left side of the barn into a 5-foot alley for driving hogs into the sales ring.

Market Driveways

The market driveways should be not less than 40 feet wide. They should be paved for heavy traffic. Aprons connecting these driveways with the truck docks should be 100 feet wide, and have a rolled gravel surface.

Parking Areas

A total of 210 parking spaces for motor vehicles should be provided in convenient, but well defined, areas on the market site. Each parking space should be 10 feet wide. Parking areas should have a gravel surface.

Arrangement of Facilities

The suggested layout for the market, which shows the arrangement of the facilities previously discussed, is shown in figure 44. Although the sales ring in the auction barn must connect with the cattle and odd-lot hog holding pens by means of appropriate alleys, the graded hog pens are not located on the yards to provide a direct flow to or from the ring.

In this layout separate truck docks are provided for hogs and for cattle. These docks, which are on opposite sides of the yards, are used both for receiving and loading out. Therefore, the flow lines for these species do not *cross* the yards as in the market shown in figure 42.

To expand or increase the size of the yards for hogs, additional pens and alleys would be constructed both on the rear and at the front of the market. Pens for graded hogs would be constructed toward the rear and pens for odd-lot hogs would be constructed at the front. By expanding in these directions, the same flow lines are maintained. To increase the size of the yards for cattle, additional pens would be constructed toward the rear of the market. However, the first step in expanding the cattle section of the market would be to divide the 3 large overflow pens into smaller pens and new pens of the larger size could be added. Bull pens would be expanded by taking over an adjacent 10 by 21 foot cattle pen.

Amount of Land Needed for Market Site

The amount of land needed for the market site is the same as that for the type 1 market (fig. 42).

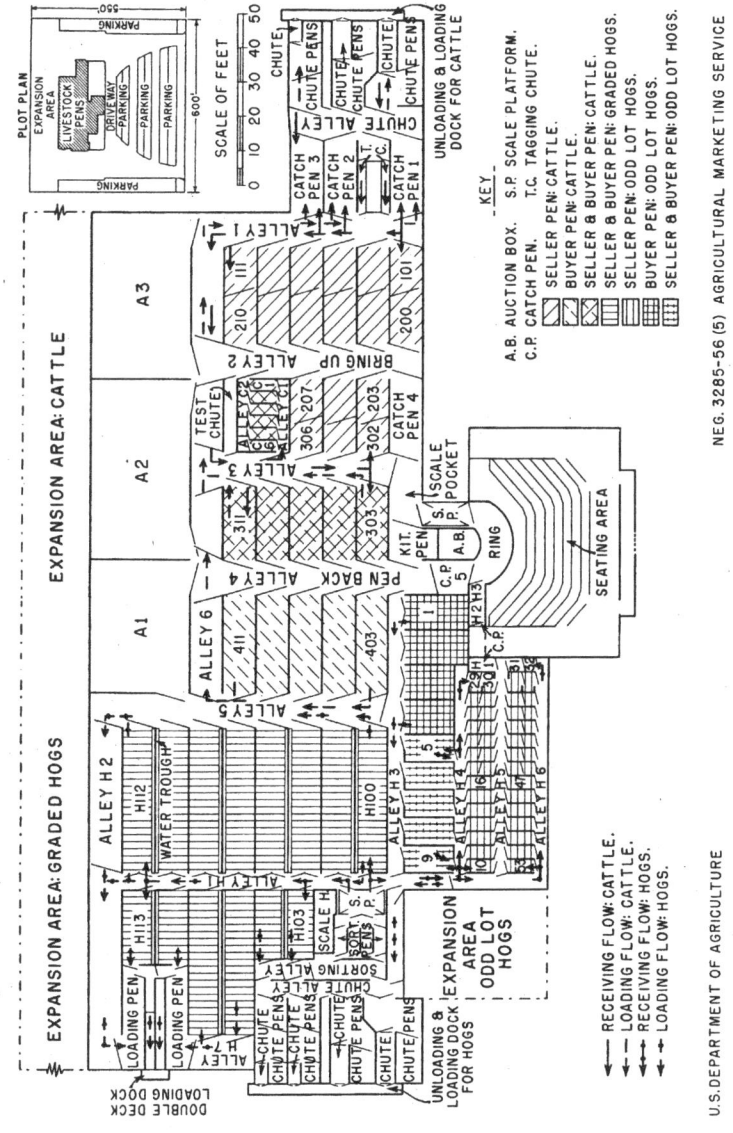

Figure 44.--Layout of the suggested market that sorts and weighs hogs on arrival and weighs cattle before their sale, with the flow of livestock during the receiving and loading cycles, and the suggested expansion area.

Estimated Cost of Construction

The basis used in determining these costs is the same as that used for determining the costs for the type 1 market. The estimates are for constructing a market to handle 1,260 hogs and 260 cattle per sale.

Item	Estimated Cost Dollars
Docks, chutes, and chute pens	1,760.00
Cattle pens:	
Fences and gates 5 ft. high, five 2- by 6-inch rails, 6- by 6-inch and 8- by 8-inch posts	
2,110 lin. ft. @ $2.40 per ft.	5,064.00
Hayracks and feed troughs	
75 lin. ft. @ $2.00 per ft.	150.00
Water troughs - 6 troughs @ $27.50 each	165.00
Paving alleys (black top) 615 sq. yds. @ $1.90 per sq. yd.	1,168.50
Hog pens:	
Fences and gates, 3 ft. 6-in. high, two 1- by 10-inch rails, six 1- by 6-inch rails, 4- by 4-inch, and 6- by 6-inch posts	
2,210 lin. ft. @ $2.65 per ft.	5,856.50
Scale, 7- by 14-ft. platform and scale house	2,750.00
Water troughs, 330 ft. @ $2.20 per ft.	726.00
Hog spray system	86.00
Paving pens and alleys (concrete) 1,390 sq. yds. @ $2.20 per sq. yd.	3,058.00
Roof over yards 30,605 sq. ft. @ $0.50 per sq. ft.	15,302.50
Catwalk over yards 600 lin. ft. @ $2.75 per ft.	1,650.00
Yard lighting	550.00
Auction barn, frame construction	
Includes stairs on front, auction box, offices, restaurant, and toilets	
3,060 sq. ft. @ $3.30 sq. ft. 1/	10,098.00
Scale, 7- by 14-ft. platform	2,420.00
Seats, 213 theatre type @ $5.50 each	1,171.50
Public address system, 2 systems @ $550 each	1,100.00
Ticket carrier system	165.00
Water lines in cattle and hog pens, cafe, and toilets	1,320.00
Sewer lines (inside market layout)	1,980.00
Paving (rolled gravel):	
Driveways, aprons, and parking	
18,100 sq. yds. @ $0.55 sq. yd.	9,955.00
Subtotal	66,496.00
Engineering fees @ 6 percent	3,989.76
Total cost of market	70,485.76

1/ Estimate includes all electrical work, lighting fixtures, plumbing, and toilet facilities but does not include office and restaurant equipment.

How Suggested Facility Should Operate

The various components of the market are arranged so that definite flow lines must be observed during the 3 major cycles of operations. These flow lines for hogs and cattle are shown in figures 44 and 45.

Receiving Operations

After hogs are unloaded, each load is driven from the chute pen across the chute alley and into the sorting alley. They are sorted by market grades in the sorting lane and driven into sorting pens. From these pens, hogs are driven one pen lot at a time onto the scale platform where they are weighed and the receiving ticket is prepared. From the scale platform hogs meeting grade specifications are driven into and down alley H 1 to the graded hog pens H 100 to H 113. Hogs not meeting the specifications of market grades and other odd-lot hogs are sorted into salable lots in the sorting pens, weighed, and are yarded through alleys H 1 and H 6 in pens H 31 to H 53 and through alleys H 1 and H 4 to pens H 10 to H 30 and H 5 to H 9. The first pens in which these hogs are yarded depends on the size of the lots. Small size lots are yarded in pens H 17 to H 46. Medium size lots are yarded in pens H 10 to H 16 and H 47 to H 53, and large size lots are yarded in pens H 5 to H 9.

With the exception of 4 pens, all graded hog pens are designed to hold about a trailer deck load. The 4 small pens are used for those grades that usually run less than a trailer deck load. Sorting into trailer dock lots in loading out is eliminated. All graded hog pens are near the *off* end of the scale platform so that hogs can be driven from the platform to any of these pens with minimum labor requirements. This arrangement also eliminates the necessity for moving hogs from one pen to another during the course of the receiving.

Cattle unloaded from large trucks are driven from the chute pen across the chute alley and into the tagging chute where they are tagged and the receiving ticket prepared. From the tagging chute cattle are driven into catch pen 1, 2, or 3. Loads arriving on pickup trucks are tagged and the receiving ticket prepared while the animals are on the truck. After unloading, these cattle are driven directly to the catch pens.

The first holding pens used for yarding sellers' cattle are dual-purpose pens 303 to 311. Cattle yarded in these pens are driven through alleys 1, 6, and 3. Cattle yarded in pens 302 to 306 and 203 to 207 are driven through alleys 1, 6, and 3. Cattle yarded in pens 200 to 210 and 101 to 111 are driven through alley 1. Catch pens are used for the last cattle arriving. Bulls are driven straight from the truck docks to the assigned pens through alleys 1, 6, 3, and C 1.

Selling Operations

Although graded hogs are sold in the sales ring, these hogs are not driven into the ring. Therefore, it is unnecessary to provide connecting alleys from their pens to the sales ring and back to the pens. However, odd-lot hogs are driven into the sales ring and provisions must be made for necessary alleys from these pens into the sales ring and back to the pens to provide a direct flow.

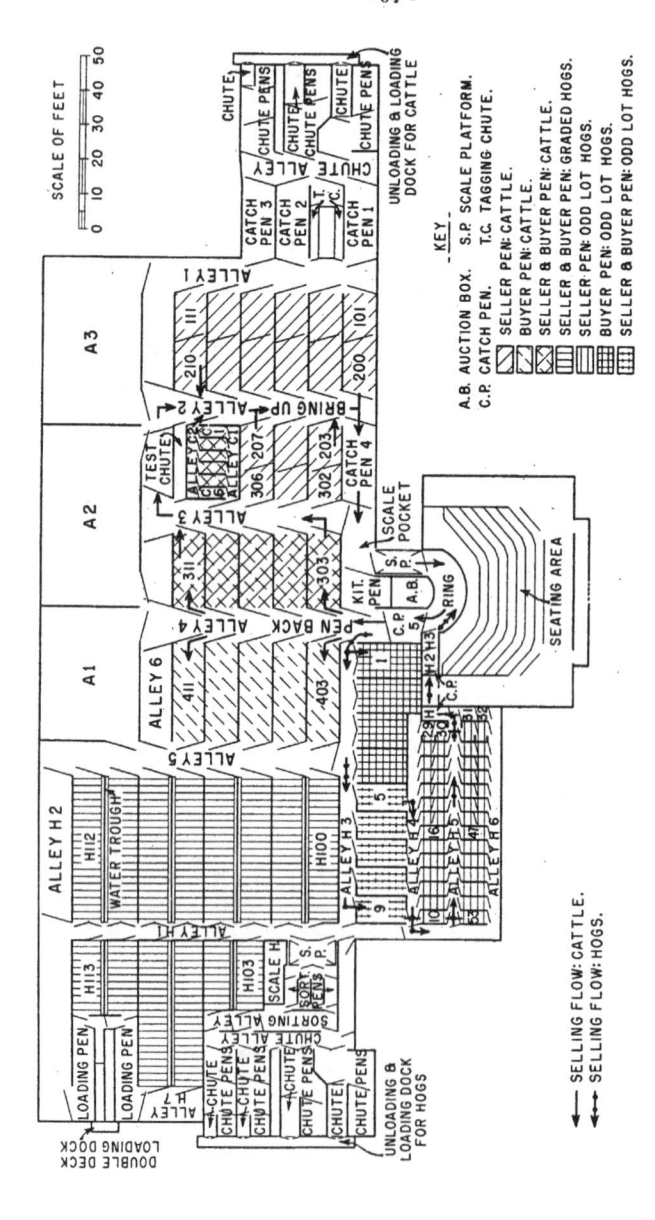

Figure 45.—The flow of cattle and hogs during the selling cycle on the suggested market that sorts and weighs hogs on arrival and weighs cattle before their sale.

Hogs yarded in dual-purpose pens H 5 to H 9 are the first to be sold so that these pens may be made available to buyers. Hogs from these pens are driven through alleys H 1 and H 5 to catch pen H 1. In catch pen H 1 hogs are sorted into salable lots and driven into catch pens H 2 and H 3 for holding adjacent to the sales ring. Odd-lot hogs yarded in pens H 10 to H 30 and pens H 31 to H 53 are driven to catch pen H 1 through alley H 5.

Odd-lot hogs are driven into the sales ring, displayed, and driven out through catch pen 5, alley 4, and alley H 3 to buyer pens H 1 to H 9. Pens H 1 to H 4, the largest buyer hog pens, are assigned to the largest buyers to minimize the distances of the drives from the sales ring to the pens.

Cattle in the dual-purpose cattle pens 303 to 311 are the first sold. From these pens cattle are driven into alley 3 through alley 6, and up alley 2 to catch pen 4. Cattle in pens 203 to 207 and 302 to 306 are driven into and up alley 2 to catch pen 4. Cattle in pens 200 to 210 and 101 to 111 also are driven to catch pen 4 through alley 2. Catch pen 4 holds about 15 cattle. Cattle are driven in small lots from catch pen 4 into the scale pocket. No more than 5 or 6 head should be driven into this pen at one time because the worker driving animals onto the scale platform can work more efficiently when the pen is not too crowded. Cattle are driven from the scale platform into the sales ring as needed.

Cattle are driven from the sales ring into catch pen 5 where they are temporarily held pending their assignment to a buyer pen. From catch pen 5 animals are driven into and down alley 4 to the assigned buyer pen.

The suggested arrangement of workers in the sales ring and the auctioneer's box is similar for hogs and cattle. In selling odd-lot hogs one ringman should be stationed at the *in* gate of the ring and the other ringman on the opposite side of the ring. He *turns* the hogs so they can be driven out of the ring through the gate next to the one where they entered. The recorder at the counter in the auctioneer's box should be next to the auctioneer on the side that animals enter the sales ring. The worker assigning hogs to buyer pens is located near the *out* gate of the ring. No weighmaster is used during the sale of hogs.

The sale would be conducted similar to the method previously described for the type 1 market. The major difference would be that the auctioneer begins the sale by submitting a bid or suggesting a price.

In selling cattle the organization of the selling crew would be about the same as that described for hogs. However, a weighmaster is added to the crew and is located in the auction box on the side near the gate through which cattle enter the sales ring and just to the rear of the recorder so he can pass the scale ticket to the recorder.

Loading-Out Operations

The physical handling involved in loading-out livestock requires fairly short and direct drives of animals in relatively large lots. Hogs are loaded out either at the same truck dock where they are received or at the double-deck dock.

Hogs in the graded pens H 100, H 101, H 102, H 104, H 106, H 108, H 110, and H 112 are driven into and down alley 5 to alley H 2 through alley H 2 to the loading pens for the double-deck dock. To load out at the other docks hogs in these pens would be driven into and up alley H 1 to alley H 3, through alley H 3 to the chute alley. Hogs yarded in pens H 111 and H 113 would be driven out of the pen directly to the loading pens of the dock when being loaded out at the double-deck truck dock. Hogs yarded in pens H 5 to H 9 would be driven into alley H 4 and up alley H 1 to alley H 3, through alley H 3 into the chute alley when they are loaded out at the truck docks at which they were received. Hogs yarded in pens H 1 to H 4 would be driven through alley H 3 to chute alley or alleys H 3, 5, and H 2 to double-deck docks. When it is necessary to load-out hogs yarded in pens H 103 and H 105 at the same docks at which they were received, they are driven directly from the pens into the chute alley, through the chute pens and onto the trucks. Hogs yarded in pens H 107 and H 109 are driven through alley H 7 to the loading pens when being loaded out at the double-deck truck docks.

In loading-out cattle, those in pens 403 to 411 are driven into and down alley 5 to alley 6, through alley 6 to alley 1, and through alley 1 to the truck docks. Cattle yarded in pens 303 to 311 are driven into and down alley 3 to alley 6, through alley 6 to alley 1, and up alley 1 to the truck docks.

Labor Requirements

Fifteen workers, other than the office crew, are required to operate the suggested market, or 3 fewer than the number required for the typical market. The reduction of workers is in the number of yardmen required and is made possible through improved design and layout of the facilities. The crew for the suggested market consists of the auctioneer, weighmaster, recorder, 2 ringmen, the yard foreman, and 9 yardmen.

The hog receiving crew is 7 workers as compared with 8 for the typical market. One worker prepares the receiving ticket, 1 worker drives hogs from the dock to the sorting pen, 2 workers sort hogs, 1 worker weighs hogs, and 2 workers yard hogs in seller pens. The worker eliminated assists in penning hogs in sellers' pens. This reduction is made possible because of the arrangement of the holding pens in relation to the sorting pen and scales, which results in a compact working area, thereby reducing the distance of travel of hogs between work stations. The 2 workers suggested for receiving cattle is one less than the number required on the typical market and is made possible through improved receiving facilities. When the sale starts 5 workers receiving hogs and 1 worker receiving cattle are shifted to jobs in the selling operation. The 3 remaining workers receive and load out both cattle and hogs during the selling operations. The yard foreman directs the activities of all workers receiving hogs and cattle.

The hog selling crew consists of 11 workers: The auctioneer, recorder, 2 ringmen, and 7 yardmen. Of these workers 6 are obtained from the receiving operations and 5 workers report for duty at the start of the sale. This number of workers is 3 fewer than the number required for the typical market. The arrangement of the hog pens in relation to the sales ring decreases the number of workers required to bring up hogs for sale and yard hogs after their sale.

The cattle selling crew consists of 12 workers: The auctioneer, weighmaster, recorder, 2 ringmen, and 7 yardmen. The yard foreman weighs the cattle. The number of workers required in selling cattle is 3 fewer than is required for the typical market. This reduction in the number of workers needed is in the yard crew. The arrangement of the holding pens, alleys, and scale platform in relation to the sales ring decreases the number of workers required to bring up and drive cattle into the sales ring by one worker. The straightaway alley from the sales ring to buyer pens and the 10-foot width buyer pens on the suggested market make it possible to yard cattle just as quickly and efficiently with 2 less workers.

The number of workers needed to check animals in the buyer pens on the suggested market is 10, or 5 fewer than required for the typical market. However, these 10 workers should be able to perform this operation on the suggested market in a shorter period because pens are accessible from either side and adequate block gates and space in alleys are available to these workers.

SUGGESTED LAYOUTS FOR MARKETS THAT WEIGH LIVESTOCK BEFORE THEIR SALE--TYPE 3

The types and amounts of facilities needed and their arrangement for markets that weighs livestock before their sale is shown for 2 subtypes of markets. One market tags but does not dip cattle. The other market dips but does not tag cattle.

Layout for Market That Tags but Does Not Dip Cattle

To show the types and amounts of facilities needed and their arrangement for a type 3 market that tags but does not dip cattle, a market handling about 600 cattle and 150 hogs per sale is selected (fig. 46). This market mixes owner lots in the same pens, provides separate buyer and seller pens, and sells most cattle singly. Livestock are received and shipped by motortrucks.

With but few exceptions, the facilities needed and their arrangement are the same for this market as for the type 1 market (fig. 42). The major differences are: (1) The feeder chute facility terminating at the entrance gate to the sales ring is eliminated and a scale platform is inserted in its place; (2) a catch pen is located adjacent to the *out* sales ring gate; and (3) a ground floor 5-foot passageway is constructed on the left side of the sales ring. The location of the scale platform on the suggested market permits animals to be weighed before they are driven into the sales ring. A catch pen 10 by 10 feet adjacent to the *out* gate of the sales ring permits animals to be held for short periods before they are assigned to pens. The ground floor passageway on the left side of the sales ring provides an alley for driving hogs from the ring to buyer pens over the shortest route.

The arrangement of cattle holding pens and alleys in the suggested market is the same as that for type 1 market. Therefore, the flow of cattle into, through, and out of the market would be the same. Although seller and buyer hog pens are in the same general area of the market yards, their arrangement is reversed. Hogs would move over the scale platform into the sales ring and from the ring to buyer pens through the 5-foot passageway at the left side of the sales ring.

Layout for Market That Dips but Does Not Tag Cattle

To show the types and amount of facilities needed and their arrangement at a market that weighs livestock before their sale and that dips but does not tag cattle, a market handling about 600 cattle and 40 hogs per sale is selected. This market provides both sellers and buyers with individual holding pens regardless of the size of lots. As neither

- 88 -

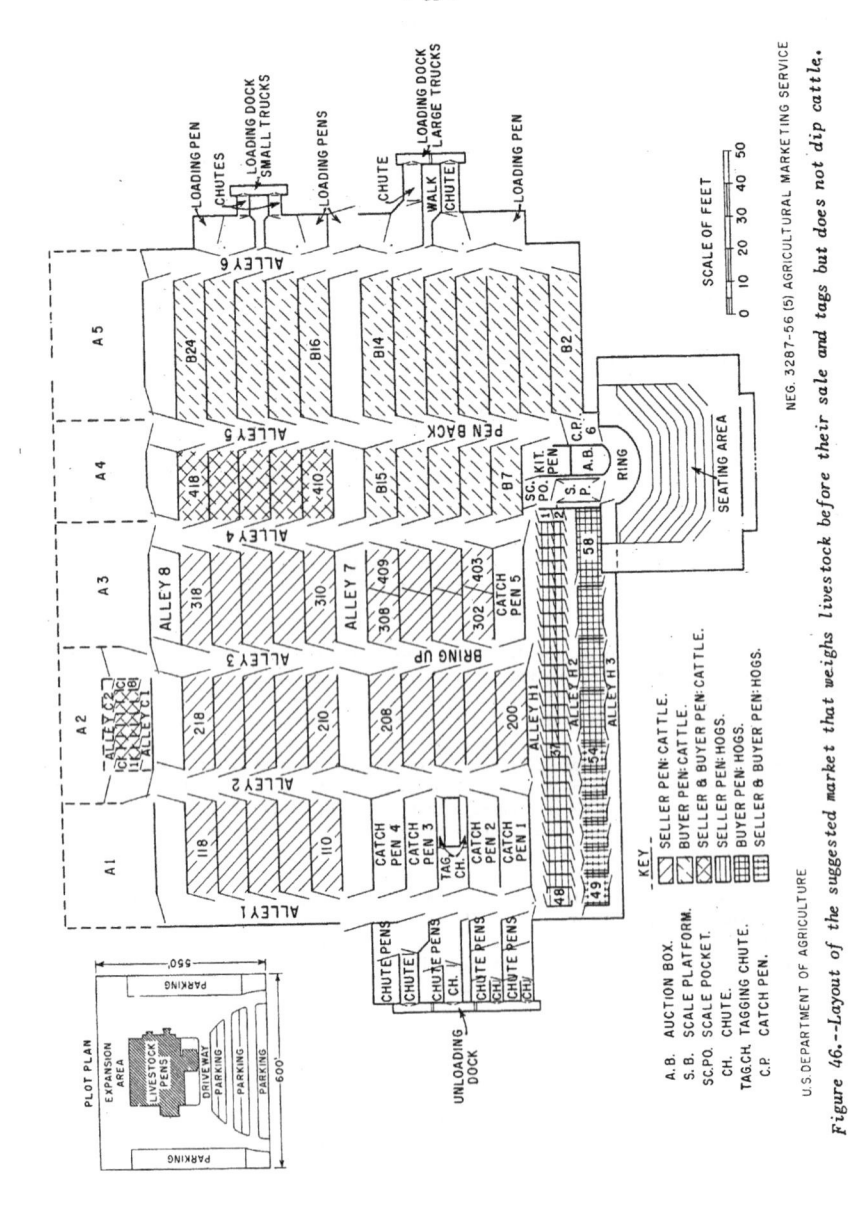

Figure 46.--*Layout of the suggested market that weighs livestock before their sale and tags but does not dip cattle.*

cattle nor hogs are tagged, owner lots are never mixed in the same pen. A large number of animals are sold singly. Livestock are received and shipped by motortruck.

Facilities Needed

Truck docks and Loading Pens.--The truck docks suggested for this market should be of the design, size, and arrangement shown in figure 28. The amount of truck dock space both for receiving and loading, and loading pen area needed for this market are the same as that described for type 1 market.

Dipping Vat.--In Florida the State Livestock Sanitary Board provides construction details for cattle dipping vats for auctions. Copies of plans may be obtained from the Board on request A 72-foot dipping vat should include a catch pen, vat, and drip pen. The catch pen and drip pen are at opposite ends of the vat. Each pen should contain about 300 square feet of space and would hold about 25 head of average size cattle.

Catch Pens, Sorting Pens, and Scale Pocket.--Two catch pens are suggested for the yards proper; both are for selling cattle. One catch pen, 10 by 38 feet, should be constructed with an interior gate so that 2 pens can be formed by closing this gate. One end of the pen connects with the major bring-up alley and the other end connects with the service alley of the sorting pens. Animals are held in the pen connecting with the bring-up alley and are sorted, when necessary, into the pen that connects with the service alley.

The service alley which serves the sorting pens is 5 by 20 feet. Four sorting pens, each 5 by 10 feet, are located between the service alley and the scale pocket. Each of the sorting pens contains 50 square feet and holds about 4 animals. The scale pocket is 5 by 20 feet and is adjacent to the scale platform. The scale pocket is used to hold animals prior to driving them onto the scale platform.

The other catch pen used for selling cattle is 10 by 14 feet and is adjacent to sale ring *out* gate. This pen is used to hold cattle temporarily pending the completion of their sale or their assignment to a buyer pen. For selling hogs, 3 sorting pens each of which is 4 by 8 feet, are located near the sales ring and a short alley adjacent to the scales.

The assumed market would handle about 600 cattle and 40 hogs per sale. Although the market might handle a few sheep, goats, horses, and mules, it would not handle a sufficient volume of these species to justify specially-designed holding pens for them. Cattle pens would be used for holding horses, mules, sheep, and goats. Specially-designed bull pens would be needed.

Markets of this type provide individual pens for each buyer and seller regardless of the size of owner lots. Owner lots are identified by pen numbers rather than by a tag number as on other markets. As a consequence, the market must have a sufficient number of holding pens to accommodate the largest number of sellers and buyers on peak sale days. This practice of providing each seller and buyer with large individual holding pens means that it is rarely possible to utilize more than 70 percent of the space in cattle pens. By varying the size of cattle pens and providing more smaller pens, the utilization of pen space could be increased to 80 percent of total capacity. For hogs, the utilization of pen space would remain about the same.

The amount of pen space needed per animal is 14 square feet for cattle and 6 square feet for hogs.

Table 5 shows the types of pens, the number of pens of each type, and the total space needed at a livestock auction that will handle about 600 cattle and 40 hogs. Included are 51 seller cattle pens. To provide for the varying size loads and numbers of sellers, 27 of the cattle pens should be constructed with middle gates which can be closed to form 54 pens. A total of 78 individual cattle pens would be available for any given sale. In addition, 10 bull pens are suggested. Thus the maximum number of individual pens available for cattle sellers would be 88.

Table 5.--Types and number of holding pens and total pen area for suggested market that practices presale weighing of livestock

Types and sizes of pens	Pens	Pen area
	Number	Square feet
Sellers' cattle pens		
10 by 42 feet 1/	11	4,620
10 by 35 feet 1/	5	1,750
10 by 28 feet 1/	11	3,080
8 by 8 feet 9 inches	8	560
4 by 8 feet	16	512
Subtotal	51	10,522
Buyers' cattle pens		
10 by 21 feet	11	2,310
10 by 42 feet	11	4,620
8 by 8 feet	4	256
6 by 8 feet	2	96
4 by 8 feet	10	320
Subtotal	38	7,602
Dual-purpose bull pens 2/		
8 by 3 feet 6 inches	10	280
Total cattle pen space		18,404
Sellers' hog pens		
4 by 4 feet	16	256
Buyers' hog pens		
8 by 8 feet	1	64
4 by 8 feet	6	192
Subtotal	7	256
Total hog pen space		512
Total holding-pen space		18,916

1/ These pens should have gates that open on 2 alleys and each pen should have a center gate which can be closed to divide the pen into 2 pens.

2/ Dual-purpose pens are pens that might be used alternately by either sellers or buyers.

A total of 38 buyer pens is suggested. Ten of these are 4 by 8 feet, 2 are 6 by 8 feet, and 4 are 8 by 8 feet. These pens are for small buyers. All bull pens would be used as dual-purpose pens and provide a total of 48 pens for buyers.

A total of 10,522 square feet of space is provided in the 51 seller cattle pens. Allowing 14 square feet per head and with 80 percent utilization of space, about 600 cattle can be accommodated, plus 10 bulls.

A total of 7,602 square feet of space is provided in the 38 buyers' cattle pens. Allowing 14 square feet per head with 80 percent utilization, 434 cattle could be held in these pens. When buyer pens were filled cattle would be penned back in empty seller pens.

The 16 seller hog pens provide 256 square feet of space. Allowing 6 square feet per head, these pens would accommodate 42 hogs. The 7 buyer hog pens provide the same amount of space and would accommodate the same number of hogs.

The description of the floors and drainage is the same as that previously discussed for type 1 market.

Alleys.--In the cattle section of the market, alleys should be 10 feet wide, except those in the bull pen and small cattle pen sections. Specifications for alleys in the cattle section are the same as those described for the type 1 market. Alleys in the hog pen sections are 4 feet wide. These alleys should be paved.

Facilities for Driving Livestock into Sales Ring.--Facilities for driving cattle into the sales ring consist of a service alley, 4 sorting pens, a scale pocket, and the scale platform. The service alley connects directly with the catch pen in which animals are held during the sale.

The sorting pens have 5-foot gates on each end--one gate opening into a service alley, and one opening into the scale pocket. The in gate of the scale platform connects with scale pocket. Therefore, cattle move directly through the sorting pens onto the scale platforms. Multiple sorting pens enable workers to sort lots of cattle at the same time other operations are being performed. Therefore, selling operations are rarely delayed by the sorting operation In areas where nervous cattle are predominant, markets should construct walks across the top of these facilities to minimize injuries to workers.

Facilities for driving hogs into the sales ring consist of 3 sorting pens and 2 service alleys. Each of the alleys are about 4 by 12 feet. The sorting pens are between the 2 alleys. The 4-foot gates at each end of each pen open into these alleys. The scale platform adjacent to the service alley has a 4-foot gate constructed in the side so that hogs can be driven directly onto it.

Scales.--The scale for weighing livestock should be the weighbeam type, having 10,000-pound capacity, a 5-pound minimum graduation, printing attachment, and 7- by 14-foot platform. The rack for scale should be of the design shown in figure 34.

Roof.--The entire market yards should be covered by a monitor-type roof. About 35,916 square feet of roofing would be needed.

Other Yard Facilities.--This market should be equipped with blockgates and lights similar to those described for the type 1 market. About 540 feet of catwalks would be needed to provide a loop over the market yards (fig. 40). Two kitchen pens 10 by 14 feet and 10 by 10 feet respectively, are necessary.

In the cattle section watering and feeding facilities should be provided in at least 5 buyer pens. Two 10-foot watering troughs and one 20-foot combination hayrack and grain trough should be provided for each of these pens, or a total of 100 feet of water troughs and 100 feet of combination troughs. (See figures 37 and 39 for suggested designs.)

In the hog section of the market one water trough running the full length of the pens should be provided in each of the buyers' hog pens, for a total of about 56 feet of troughs. In addition, adequate hydrants with threaded hose connections also should be provided in the hog section for wetting down and washing out pens (fig. 38).

Auction Barn.--This auction barn is similar to that described for type 2 market except that the 5-foot alley is used for driving hogs out of the sales ring rather than into it (fig. 41).

Market driveways and parking areas are also about the same as those for market 2.

Arrangement of Facilities

The suggested layout for the market is shown in figure 47. The auction barn fronts on and is connected with a public highway by a 40-foot paved driveway. The manner in which this driveway connects with 100-foot service driveways on each side of the market is similar to that previously described. The number and arrangement of parking spaces for vehicles also is about the same.

The sales ring of the auction barn connects with the 2 drive alleys in the market yards used respectively for moving cattle up to the ring for sale and penning back after the sale. Catch pen 2, the sorting pens, a scale pocket, and the scale platform pen (rack) provide this connection from the incoming alley. Catch pen 3 provides the connection between the ring and the outgoing alley. The sales ring also connects with 2 drive alleys used respectively for moving hogs up to the ring for sale and penning back after the sale. The scale platform pen (rack) provides the connection for incoming animals and a gate in the sale ring provides the connection for outgoing alley. The seating area and auctioneer's box also connects with the yards by means of steps from the catwalks.

Sellers' cattle holding pens are grouped in 3 rows along the left side of the market yards (the left side when the viewer faces the yards from the auction barn), the side on which livestock would be received. Four rows of small-size cattle sellers' pens and one row of bull pens are located near the front of the yards in the center row of sellers' pens. Buyer cattle holding pens are grouped in 2 rows on the opposite side of the yards where cattle would be loaded out. Large size buyer pens are nearest the sales ring and small size buyer pens are at the rear of the outside row of buyer pens.

Hog holding pens are grouped in 2 rows at the front of the market near the sales ring. The inside row is seller pens and the outside row is buyer pens.

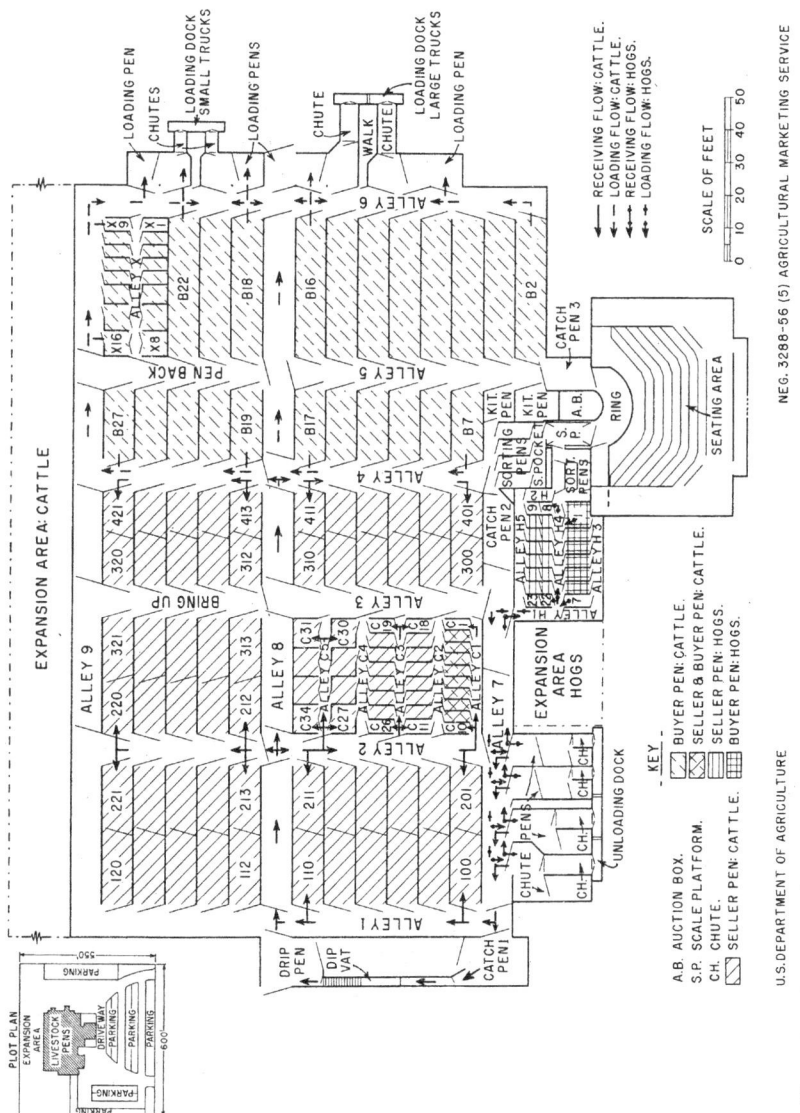

Figure 47.--*Layout of the suggested market that weighs livestock before their sale and dips but does not tag cattle, with the flow of livestock during the receiving and loading cycles, and the suggested expansion area.*

The large size cattle pens in the 5 rows are separated by and open into alleys running from the rear to the front of the yards. The 4 interior alleys separating these rows of cattle pens are parallel to alleys on each of the outer edges of yards so that each of the large size pens open on each side, into 2 different alleys. Cross alleys are provided at the front, rear, and about midway of the yards.

The 4 rows of small size cattle seller pens and 1 row of bull pens are separated by small cross alleys that open into major drive alleys. The bull pens and 3 rows of small size cattle pens open onto 2 alleys, 1 row of the small size pens opens into only 1 alley.

With the exception of alleys in the bull pen and small seller and buyer cattle pen sections all alleys in the cattle section of the yards are 10 feet wide. Alleys in the bull pen and small seller and buyer cattle pen section are 4 feet wide. Full width gates on all pens serve as blockgates for alleys. Drive alleys are numbered from 1 to 6, inclusive, beginning at the left side of the yards. Cross alleys are numbered 7, 8, and 9 beginning at the front of the yards. Alley number 3 is the main drive alley used to bring up cattle to the ring. Alley number 5 is the pen back alley.

Two rows of hog pens are separated by alleys and all hog pens open onto 2 alleys. Alleys 2 and 5 are the main drive alleys for bringing up hogs to the sales ring. Alley H 3 is the pen back alley. Alley H 1 is the only cross alley in the hog section.

Truck docks for unloading livestock are on the front of the yards at the left side of the auction barn. This location provides for a free flow of animals to the dipping vat which is connected to the yards by alley 7 and catch pen 1. The drive *out* end of the vat is connected to the yards by the drip pen and alley 8. A free flow of cattle is provided by this arrangement.

The 2 truck docks for loading-out livestock should be on the right side of the yards. Loading pens are adjacent to each of the 4 spaces on the 2 docks for loading out. Each pen opens onto the same drive alley.

The scale platform is located between the *in* gate of the sales ring and the scale pocket so that one scale gate forms part of the sales ring and the other opens into the scale pocket. This location places the scale alongside the auction box and permits the scale beam to be placed inside the box. This location also permits the workers driving livestock around the ring to open and close the gate of the scale platform.

To expand or increase the size of the yards, additional pens and alleys for cattle would be constructed to the rear of the yards. Pens constructed to the rear of the market should be large size pens and the same flow lines maintained. If smaller pens are needed it is suggested that large pens 212 to 220 and 313 to 321 be divided into small pens. The arrangement of these pens should be similar to that of the small pens now shown in the layout. Hog holding pens would be expanded by dividing alley H 1 into pens and adding additional pens to the rear of the present pens. Alley H 1 should be relocated to the rear of the expanded pens in the same manner as now shown in the layout.

Estimated Cost of Construction

Cost of construction estimates for the type 3 market providing separate buyer and seller pens and dipping cattle are shown below. This market is designed to handle about 600 cattle and 40 hogs per sale.

Item	Estimated Cost Dollars
Docks, chutes, and chute pens	1,760.00
Cattle pens:	
Fences and gates 5 ft. high, five 2- by 6-inch rails,	
5,440 lin. ft. @ $2.40 per ft.	13,056.00
Hayracks and feed troughs	
100 lin. ft. @ $2.00 per ft.	200.00
Water troughs	
10 troughs @ $27.50 each	275.00
Dipping vat (concrete)	1,650.00
Paving alleys (black top)	
1,550 sq. yds. @ $1.90 per sq. yd.	2,945.00
Hog pens:	
Fences and gates, 3 ft. 6 in. high, two 1- by 10-inch rails	
six 1- by 6-inch rails, 4- by 4-inch, and 6- by 6-inch posts	
424 lin. ft. @ $2.65 per ft.	1,123.60
Water trough	
56 lin. ft. @ $2.20 per ft.	123.20
Paving pens and alleys (concrete)	
137 sq. yds. @ $2.20 per sq. yd.	301.40
Roof over yards	
35,916 sq. ft. @ $0.50 per sq. ft.	17,958.00
Catwalks over yards, including 2 stairways	
540 lin. ft. @ $2.75 per ft.	1,485.00
Yard lighting	550.00
Auction barn, wood frame construction 1/	
Includes stairs on front, auction box, offices, restaurant, and toilets	
3,060 sq. ft. @ $3.30 per sq. ft.	10,098.00
Scale 7- by 14-ft. platform	2,420.00
Seats, 213 theatre type @ $5.50 each	1,171.50
Public address system, 2 systems @ $550 each	1,100.00
Ticket carrier system	165.00
Water lines in cattle and hog pens, cafe, and toilets	1,320.00
Sanitary sewers (inside market area)	1,980.00
Paving (rolled gravel):	
Driveways, aprons, and parking (gravel)	
18,100 sq. yds. @ $0.55 per sq. yd.	9,955.00
Subtotal	69,636.70
Engineering fees @ 6 percent	4,178.20
Total cost of market	73,814.90

1/ Estimate includes all electrical work, lighting fixtures, plumbing and toilet facilities, but does not include office and restaurant equipment.

Amount of Land Needed for Market Site

The amount of land needed for the market site is the same as that for the type 1 market.

How Suggested Facility Should Operate

The flow lines that must be observed during the 3 major cycles of operation, if the facility is to operate efficiency, are shown in figures 47 and 48. Other flow lines would be possible on this market by swinging some of the gates differently.

Receiving Operations.--Each consignment of cattle received at the unloading docks is driven from the chute pen into and through alley 7 to catch pen 1 at the head of the dipping vat. Each animal then is driven single file into and out of the vat to the drip pen where the entire lot is collected and it is allowed to stand for a short period while the dipping solution drips off. From the drip pen, each lot is yarded in seller pens. Small consignments are yarded in pens C 1 to C 34, located about half way between the drip pen and the sales ring, through alleys 8, 2, and C 1, 3, and 5.

Larger consignments are yarded in pens 312 to 320 and 413 to 421 through alleys 8 and 4. Cattle yarded in pens 300 to 310 and 401 to 411 are driven through alleys 8 and 4. Cattle yarded in pens 100 to 120, 201 to 221, 212 to 220 and 313 to 321 are driven through alleys 8 and 2.

Each consignment of hogs is driven from the chute pen at the truck dock into and through alleys 7, H 1, and H 4 to seller hogs pens H 8 to H 23.

Selling Operations.--The movement of animals during the sale is shown in figure 48. Sales operations require the most extensive movement of animals of all market handling. All operations in this cycle must be keyed to the rate of selling.

Hogs are the first species sold. Hogs yarded in seller pens H 9 lead off followed by those yarded in pens H 8 to H 23. Hogs are driven out of the pens into alley H 5, through this alley and alley H 2 to the small service alley adjacent to the sorting pens. They are sorted in the service alley into the sorting pens. From the sorting pens hogs are driven into the scale pocket, then onto the scale platform and, after they are weighed, into the sales rings. After their sale hogs are driven through alley H 3 to the assigned buyer pens.

Cattle are sold in the same order that they arrive on the market. As the size of consignments arriving on the market are not uniform, the cattle sale may draw lots intermittently from different pen areas in the market. In general the flow of cattle from seller pens is the same regardless of the sequence in which different pen lots are brought up.

Cattle brought up from pens C 1 to C 34 are driven through alleys C 2, C 4, or C 5 into alley 3; the main bring-up alley, and through alley 3 to catch pen 2. Cattle yarded in pens 300 to 320, 313 to 321 and 212 to 220 are driven into and through alley 3 to catch pen 2. Cattle yarded in seller holding pens 401 to 421 are driven through alleys 4 and 3 to catch pen 2. Many of the cattle yarded in these pens might be driven through adjacent pens, if the latter pens were emptied first, into alley 3 and up alley 3 to catch pen 2. Cattle yarded in pens 201 to 221 are driven through alleys 2 and 3 to catch pen 2. Cattle yarded

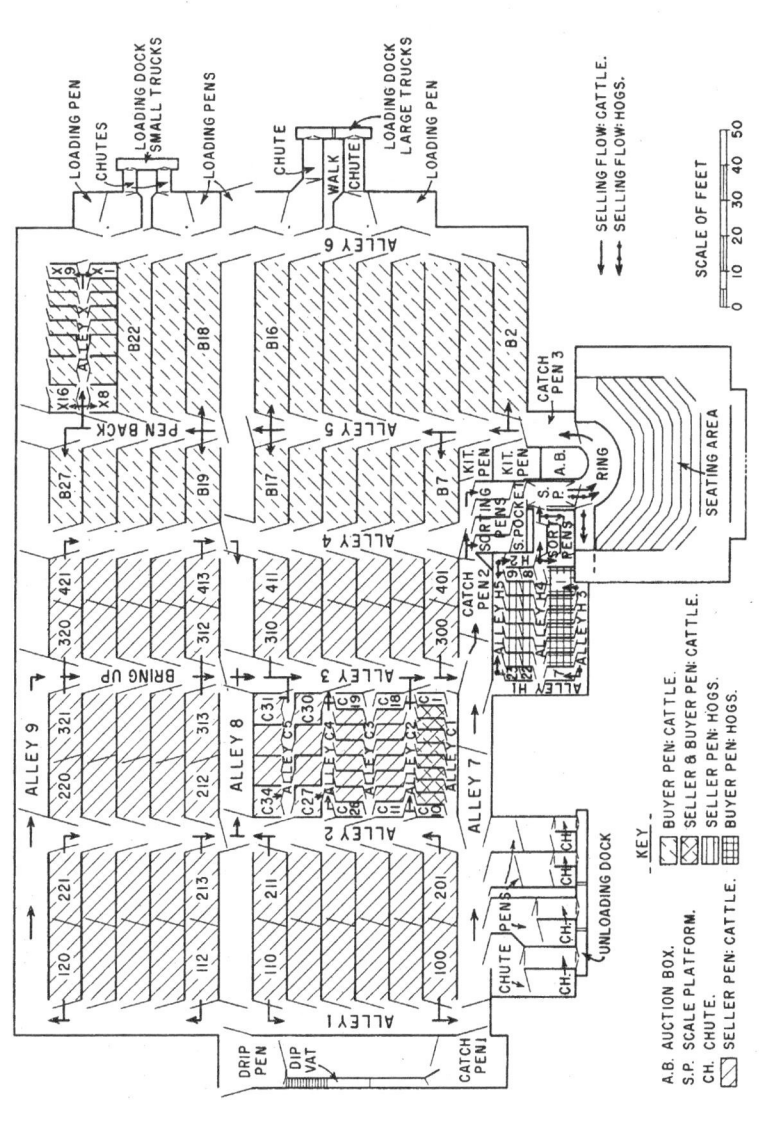

Figure 48.--The flow of cattle and hogs during the selling cycle on the suggested market that weighs livestock before their sale and dips but does not tag cattle.

in pens 112 to 120 are driven through alleys 1, 9, and 3, and cattle yarded in pens 100 to 110 are driven through alleys 1 and 7 to catch pen 2.

Cattle are sorted into salable lots from the outbound *end* of catch pen 2 into the service alley and connecting sorting pens. The sorting operation is irregular on most markets and is performed only at the owners request. If the animals are not sorted they are driven from catch pen 2 into the service alley and sorting pens in the same manner in which they were consigned to the market. From the sorting pens animals are driven into the scale pocket, then onto the scale platform for weighing, and from that point into the sales ring. After their sale cattle are driven into catch pen 3, then through alley 5 or alley 5 and alley X to buyer pens. Alley 5 is the main pen back alley. Alley X also is used for yarding cattle into pens X 1 to X 16, which are assigned to small purchasers.

A major consideration in obtaining efficiency in sales operations is the proper arrangement of workers in the sales ring and the auctioneer's box. The principles outlined for selling cattle in the type 2 market would be applicable to this market.

Loading-Out Operations.--The flow diagram of the loading-out operations is shown in figure 47. Loading-out operations require fairly short and direct drives of animals in relatively large lots.

Hogs yarded in pens H 1 to H 7 are driven through alleys H 4 and H 1 and alley 7 to the truck docks at which they were received.

Cattle yarded in pens B 2 to B 22 are driven into and through alley 6 to the truck docks for loading-out cattle. Cattle yarded in pens X 1 to X 8 are driven into alley X, through alleys X, 9, and 6 to the truck docks. Animals yarded in penx X 9 to X 16 are driven through alleys 9 and 6 to the truck docks. Cattle yarded in pens B 7 to B 17 are driven to the docks by way of alleys 4, 8, and 6 while animals yarded in pens B 19 to B 27 are driven to the dock through alleys 4, 9, and 6.

Labor Requirements

Twenty workers, other than the office crew, are required to operate the suggested market, or 3 fewer than the number required on the typical market. The reduction in the number of workers needed is in the yardmen. The crew for the suggested market consists of the auctioneer, weighmaster, recorder, 2 ringmen, the yard foreman, and 14 yardmen.

The receiving crew for the suggested market is comprised of 8 workers as compared with 10 workers on the typical market. The reduction in the number of workers is made possible by the improved arrangement of the dock, dipping vat, and seller holding pens. One worker in the receiving crew prepares the receiving ticket, 2 workers dip the animals, and 5 workers assist in unloading and driving animals to the dipping vat and driving animals from the dipping vat to the seller pens. At the start of the sale 5 workers are shifted to jobs in the selling operations.

Sixteen workers comprise the selling crew, or 3 fewer than the number required on the typical market. Five of these workers are obtained from the receiving crew and 11 report for duty at the start of the sale. The crew for the suggested market consists of the auctioneer, recorder, weighmaster, 2 ringmen, and 11 yardmen. The design and arrangement of the facilities for sorting livestock and the relation of the scale platform to the sales ring makes it possible for 4 workers to drive animals into the sales ring. Six workers perform this job on the typical market. The *straightaway* alley from the sales ring to the buyer pens and the 10-foot width buyer pens on the suggested market make it possible to yard animals after the sale just as quickly and efficiently with 1 less worker. The number of workers needed to check animals in the buyer pens and assist in loading-out livestock immediately after the sale on the improved market is 10, or 2 fewer than is used on the typical market, because pens are accessible on each side and adequate block gates and space in alleys are available.

POTENTIAL BENEFITS FROM IMPROVED LIVESTOCK AUCTION MARKET FACILITIES

It is estimated that each of the suggested market facilities for the 3 major types of markets in the Southeast could be operated with at least 3 fewer workers than are now required on typical markets handling comparable volumes of livestock. All the workers eliminated would be yardmen. In addition, some reduction in the labor required for checking animals in buyer pens after the sale should be possible. Assuming that the 3 workers eliminated worked 10 hours a day on the typical market, the minimum for most yardmen, the saving in labor on sales days on the suggested markets would amount to 30 man-hours. Possible savings in labor in checking operations are estimated at 10 man-hours. Thus, each of the suggested markets should require 40 fewer man-hours of labor per sales day than the typical market of comparable size. On the basis of an assumed wage rate of $1 per hour, the labor savings would amount to $40 per sale or, with 50 sales annually, an annual saving of $2,000.

Losses from shrinkage, bruises and injuries, and deaths also should be reduced by the improved facilities. Less delay in unloading from trucks and shorter drives through proper-width alleys into, through, and out of the market should reduce materially the shrinkage losses. Fewer right-angle turns should minimize animals slipping and decrease losses from bruises and injuries. Properly designed and located spray systems over hog pens should eliminate death losses from overheating. Although the potential benefits incurred by a reduction in these losses are not measurable, the benefits or savings should be at least as great as, if not greater than, the benefits which should be derived from operating with a smaller labor force.

In addition to these benefits the suggested facilities should: (1) Reduce the amount of time sellers' trucks spend waiting to unload, (2) increase the rate at which buyers' trucks can be loaded--an important consideration with most buyers and an inducement for them to deal with the market, (3) assure an uninterrupted flow of livestock into the sales ring, (4) make possible the assignment of more separate pens for individual consignors with a more efficient utilization of pen space, a service that is often requested but difficult to give on some existing markets, (5) make the recruiting of labor less difficult and permit its closer supervision, (6) provide sanitary conditions which should minimize the spread of livestock diseases, and (7) provide more comfortable and modern conveniences for market patrons. All of these items should assist markets in maintaining present volumes of business and enable them to obtain new business.